ROUTLEDGE LIBRARY EDITIONS:
LANGUAGE AND LITERATURE OF THE
MIDDLE EAST

Volume 7

IBN AL HAYTHAM'S
ON THE CONFIGURATION OF THE WORLD

IBN AL HAYTHAM'S
ON THE CONFIGURATION OF THE WORLD

Edited by
Y. TZVI LANGERMANN

Routledge
Taylor & Francis Group
LONDON AND NEW YORK

First published in 1990 by Garland Publishing, Inc.

This edition first published in 2017
by Routledge
4 Park Square, Milton Park, Abingdon, Oxon OX14 4RN
605 Third Avenue, New York, NY 10017

Routledge is an imprint of the Taylor & Francis Group, an informa business

British Library Cataloguing in Publication Data
A catalogue record for this book is available from the British Library

ISBN: 978-1-138-68297-9 (Set)
ISBN: 978-1-315-45973-8 (Set) (ebk)
ISBN: 978-1-138-69883-3 (Volume 7) (hbk)
ISBN: 978-1-315-51261-7 (Volume 7) (ebk)

Publisher's Note
The publisher has gone to great lengths to ensure the quality of this reprint but points out that some imperfections in the original copies may be apparent.

Disclaimer
The publisher has made every effort to trace copyright holders and would welcome correspondence from those they have been unable to trace.

Ibn al Haytham's

On the Configuration of the World

Y. Tzvi Langermann

Garland Publishing, Inc.
New York & London
1990

Library of Congress Cataloging-in-Publication Data

Alhazen, 965–1039.
[Maqālah fī hay'at al-'ālam. English]
Ibn al-Haytham's on the configuration of the world /
edited and translated from the Arabic by Y. Tzvi Langermann.
p. cm. — (Harvard dissertations in the history of science)
Translation of: Maqālah fī hay'at al-'ālam.
Includes bibliographical references.
ISBN 0-8240-0041-2 (alk. paper)
1. Astronomy, Medieval. 2. Astronomy, Greek. 3. Cosmology—History.
I. Langermann, Y. Tzvi. II. Title. III. Title: Configuration of the world.
IV. Title: Maqālah fī hay'at al-alam. V. Series.
QB26.A4413 1990
520—dc20 90-3248

Printed on acid-free, 250-year-life paper

Manufactured on the United States of America

Table of Contents

PREFACE

Circumstances allow me to prepare only the briefest of prefaces to this welcome reprint of my book, which appeared in 1990 and is based largely on the dissertation that was approved in 1979.

The two critical developments directly related to Ibn al-Haytham's *On the Configuration of the World* are Roshdi Rashed's challenge to its authenticity as a work written by the famous Ibn al-Haytham, he who wrote the masterpiece on optics and a slew of important writings in the exact sciences and philosophy, and the information available concerning the Latin translations of *On the Configuration.*

Rashed expressed his doubts at the beginning of his *Les Mathématiques infinitésimales du IX e au XI e siècle*, first published in 1993 (though it appears not to have been released until 1996) by the Al-Furqān Islamic Heritage Foundation in London. His contentions spurred A.I. Sabra—the late and lamented scholar under whose direction I wrote the dissertation which lies at the heart of this book—to publish a two part "exercise" on Ibn al-Haytham and his oeuvre: "One Ibn al-Haytham or Two? An exercise in reading the bio-bibliographical sources," *Zeitschrift für Geschichte der arabisch-islamischen Wissenschaften* 12 (1998),1-50, and "One Ibn al-Haytham or Two? Conclusion," *Zeitschrift für Geschichte der arabisch-islamischen Wissenschaften* 15 (2002/3), 95 -108. Pages 19-22 of the first installment discuss the text here published, *On the Configuration.* On page 100 of the second part, Sabra voices his opinion that "Rashed's hypothesis is strained, unnecessary, and raises more questions than it claims to have answered." Rashed replied to Sabra's strictures a few years later in "The Configuration of the universe: A book by al-Ḥasan ibn al-Haytham?" *Revue d'histoire des sciences* 60 (2007), 47-63.

It seems to me that Professor Rashed's rejection of *On the Configuration* (the text published in this volume) is intimately connected to his pioneering work on another book by Ibn al-Haytham, *On the Configuration of the Motions of Each of the Seven Planets.* I have never fully fathomed why Ibn al-Haytham could not have written both, as indeed, he authored many works, and expressed different opinions, over the course of his career. His authorship of *On the Configuration of the World*, about which I have no doubt, does not detract in the least from the splendor of the ingenious innovations which he reveals in the books so brilliantly studied by Rashed (for his study in English see Roshdi Rashed, "The celestial kinematics of Ibn al-Haytham," *Arabic sciences and philosophy* 17 (2007), 7-55). Though my research into Ibn al-Haytham's life and works does not even approach the efforts of Professors Sabra and Rashed, I do second my mentor's feeling that the disattribution of the text published here to Ibn al-Haytham is unnecessary. In any event, Taro Mimura (recently appointed to a position in history of science at Hiroshima University) has discovered a missing part of *On the Configuration of the Motions of Each of the Seven Planets*; he informs me by private communication that Rashed's *On the Configuration* is really part I of a book called *Rectification of Astronomical Operations*. Mimura's work on that text, some of which, I believe, is being done together with Jamil Ragep, should fairly well put Rashed's arguments in favor of two Ibn al-Haytham's to rest. There seems to me no reason doubt that Ibn al-Haytham, *sāhib al-Manāzir*, is the author of the *On the Configuration of the World*.

In the dissertation I did little more with the Latin versions than to list the variants I could discern in the materials available to me at the time. Now we have a full study in the as yet unpublished dissertation of Dirk Grupe (written under the direction of Charles Burnett), *The Latin Reception of Arabic Astronomy and Cosmology in Mid-Twelfth-Century Antioch: The Liber Mamonis and the Dresden Almagest*, University of London, 2013.

Finally, there are some important additions to the already very long list of Hebrew translations and manuscript copies. A fifth (!) translation into Hebrew, different from the four others previously known to exist, is found in St Petersburg, Institute of Oriental Studies of the Russian Academy C76, ff. 88a-106b; the author's introduction is lacking but otherwise the text is complete. Moreover, a Hebrew paraphrase of two chapters is found in Moscow, Russian State Library, Ms. Guenzburg 1312, ff. [43]a-[44]b (the brackets indicate that there is no pagination in the manuscript); the chapters are an "abbreviated" treatment of the ninth orb, and the one on the fixed stars. Other manuscript copies to be added to the list are: St Petersburg, Institute of Oriental Studies of the Russian Academy B70, ff. 8a-35b (which has particularly well-drawn figures); Rome, Casanatense 3082, ff. 1a-27b; and New York, Jewish Theological Seminary of America, MS 2625, ff. 1b-32a.

Y. Tzvi Langermann
Bar Ilan University
March 2016

PREFACE

The book which I now present is a thorough reworking of my doctoral dissertation. In addition to the attempt to correct errors, refine language, and update discussions, two very substantial revisions have been incorporated: (1) both the Arabic edition and the English translation have been divided into consecutively numbered passages, allowing easy reference within each text and from original to translation; (2) the cumbersome and rambling introduction and commentary of the dissertation have been combined into a more compact and, I trust, more sharply focused introduction.

My chief argument, or thesis, is that Ibn al-Haytham's *On the Configuration of the World* is a non-technical exposé of basic astronomical teachings; it was written in particular for those whose main interests were in the areas of philosophy and natural science and who, accordingly, had a particular interest in relating the mathematical devices employed by professional astronomers to the heavenly bodies mentioned in the philosophical literature. I refute previous assertions that *On the Configuration* represents a mere reworking of Ptolemy's *Planetary Hypotheses* or that Ibn al-Haytham proposed in this work to resolve any of the outstanding problems of medieval cosmology.

This, I say, is the central thesis of the present study; however, its exposition, substantiation, and defense are *not* the main motivations or justifications for its publication. Rather, the primary reason why this book is being offered to the scholarly public is that, in my opinion, the most pressing need in the history of science, especially the history of premodern science, is the publication in a form accessible to the widest audience of the outstanding *texts* . For the plain fact is that many key texts - items whose importance has already been registered in bibliographical notices, scholarly articles, and monographs - remain inaccessible. Anyone interested in Greek thought can readily obtain a copy of Aristotle and Plato; and the historian of modern science can in most cases easily find the requisite books and journals. However, the scholar who has an interest in medieval science is not so fortunate ; some of his most important source material is still in manuscript, perhaps also in a language with which he is unfamiliar. (And still, we must not ignore the value of textual research for other periods as well: fragments of the classical heritage are continually being recovered from medieval manuscripts; and scholars of the modern period are well aware of the importance of the unpublished papers of the figures whom they study). Clearly the primary task of historians of medieval science is to make this material available to the scholarly public, publishing either full editions and translations or descriptions and summaries, according to his or her judgment of the text's significance, so that, perhaps, a later generation may attempt a meaningful synthesis.

So much for the particular needs of the history of medieval science. But the publication of our text and others of equal importance is no less significant when viewed from the broader perspective of the history of science as a field of study, and also the sister fields of the history of philosophy and the history of ideas, for at least two reasons. First of all, textual studies force us to penetrate the nitty-gritty details of the scientific concepts themselves, something which is often quite bothersome and troublesome to the perpetrators of grandiose speculations. Surely we all appreciate the value of organizing the results of our diverse and specialized inquiries into meaningful generalizations. But in the history of science we can no more afford to ignore matters of detail than in other areas of serious inquiry. The history of human thought and, in particular, the history of science, has as one of its recurring themes the story of the stubborn fact that could not be steamrolled by the big theory, the pesky details that could not be circumvented by means of pretentious, inane neologisms. These details can emerge only from a close look at the source material itself; it is also by analysis of the texts that we can hope to understand the specific contexts within which certain ideas were elaborated, thereby not only shedding light on history but also, hopefully, providing us all with the necesary caution that, in the final analysis, the tools of analysis *do* allow the separation of substantive argumentation from mere pandering and ideological kow-towing.

The second reason concerns what I hope is the ambition of all of us in the field, that is, to be able, perhaps only at some future date, to arrive at some meaningful generalizations (but not "laws") of universal relevance, i.e., which apply to all of humanity. For thinking is the distinctive activity of the human race, of which scientific thinking forms one special branch. Yet so much of what is written concerning "science" in general derives from, and , therefore, may be applicable only to, modern Western science. Modern Western civilization has indeed provided the locus for the most outstanding advances in science; but the history of science is not merely the record of the grandest breakthroughs but, rather, it is the story of humanity's age-old quest for the most precise and solid understanding of the world in which it finds itself. In the attempt at assessing this quest we have in the past and may still in the present not be completely free of ethnocentric and other-centric orientations. Clearly, however, language is in any case a serious barrier in the way of any effort at an understanding of the widest cultural and temporal scope . There is, therefore, a need to make the principal textual sources available in that language knowledge of which appears to enjoy the widest diffusion; in our time, that language surely is English. It is within this frame of mind that we regard it as important to present here a scientific text written originally in Arabic, in the Middle East, and in the Middle Ages, and to note carefully the medieval European translations into Hebrew and Latin; we view this as a small contribution to the assembly of a large library of the significant works which the various peoples and ages have contributed to the scientific enterprise.

A little over ten years separate the dissertation and the book. I shall not trouble the reader with the story of my own trials and tribulations, even if such recitals are not uncommon in the medieval literature which I love so much; I shall, however, proclaim my indebtedness to those individuals whose assistance was so crucial, though, of course, they bear no responsibility for the shortcomings of this book. I acknowledge with gratitude the help of Professors A. I. Sabra and Owen Gingerich, both of whom assisted in the original dissertation and in the present publication. Professor Bernard R. Goldstein offered some invaluable suggestions and corrections. I have profited from useful discussions with Dr. F. J. Ragep; and I wish to extend a special thanks to Dr. William Newman for his timely assistance. Dr. Avraham Lavie and Mr. Aḥmad Fatḥi both offered indispensable stylistic criticisms of the preface that I wrote in Arabic. I am most thankful to Garland Publishing for making the publication of this book possible. The Institute of Microfilmed Hebrew Manuscripts at Jerusalem very kindly granted me an extended leave of absence, part of which was spent preparing this book for publication. I must also resoundingly announce the help given by father, Mr. P. H. Langermann: spiritual assistance which was offered in the time honored Jewish tradition of the love of learning, and material aid that was offered in frank recognition of some of the noncerebral, ungentle aspects of the academic profession.

Early in the winter of 1990 I returned to Jerusalem, where I made some final corrections and revisions. There, on the eighth of January, along with many other friends, students, and colleagues, I paid final respects to one of the greatest scholars of this century, Shlomo Pines. In this as in so many other projects Professor Pines graced me with the benefit of his counsel; I will always cherish memories of the unmatched breadth of knowledge, youthful enthusiasm, and grandfatherly concern which he displayed in our many discussions. I was very much looking forward to presenting Professor Pines with a copy of this book. This is now impossible. I shall, however, dedicate it to his memory.

INTRODUCTION

The *Maqâlah fî hay'at al-ᶜâlam*, or *Treatise on the Configuration of the World*, of Ibn al-Haytham (965–ca. 1040) is an important treatise by an important scientist. The text played a fairly significant role in certain aspects of the intellectual history of the Middle Ages, appearing in three scholarly languages: Arabic, Hebrew, and Latin. A Latin translation of this work [Millás Vallicrosa, 1942] and a German synopsis [Kohl, 1923] have been published. Until now, however, the Arabic text has not been edited, nor has the entire text been given a modern translation. The Hebrew translations, which survive in a particularly large number of manuscripts, have not been studied. None of the translations has been systematically checked against the original.

We have in this book taken the steps necessary for making this text accessible. We have produced an edition of the Arabic original, and we have translated the entire text into English. The text has been checked, word by word, against the Hebrew and Latin translations, and the variants listed.

Our study is mainly concerned with the text, in accordance with our view that the immediate task of historians of Arabic science is to publish the important texts which are as yet unavailable to most of the scholarly public. In this task it would be an unnecessary burden to have to include a full biographical study of the author and a full survey of medieval astronomical theory. Fortunately, we are able to refer the reader to excellent studies of both of these matters. For biographical and bibliographical information on Ibn al-Haytham, we recommend A. I. Sabra [1972], and for all questions pertaining to pre-modern astronomy, O. Neugebauer [1975].

We have limited our remarks in the introduction to a number of discrete topics, all of which revolve about questions of cosmology. In the Middle Ages the description of the heavens was a matter of concern to at least two distinct groups. Astronomers produced geometrical models by means of which the motions of the planets could be accurately accounted for in the face of the available observational data. Natural philosophers (that is, those who elaborated a philosophy of nature; they are often called "physicists") were concerned with the qualitative, three dimensional description of the heavens and, in particular, they demanded that any such description conform to a given set of rules governing the type of structures and motions that were allowed for the celestial regions.

This neat distinction of interest along professional lines did not always hold up in practice. It was only natural that individuals versed both in astronomy and philosophy compared the two descriptions and, when they did, they found serious discrepancies. The result was a serious dilemma for those medieval thinkers who were interested in cosmology. Now Ibn al-Haytham, in the introduction to *On the Configuration*, states that one of the goals of his book is to describe the motions studied by astronomers in terms of three-dimensional bodies.

Given this state of affairs in medieval intellectual history, and given as well the unavailability of our text to modern scholars, a perhaps not surprising misconception has developed that in our text, Ibn al-Haytham attempted some solution to the cosmological dilemma. This is incorrect. Not that the problem was not of interest to Ibn al-Haytham: in another work, *Doubts concerning Ptolemy*, this eminent scientist has a lot to say on the matter. But *On the Configuration* is a text of a different sort. The correct characterization of *On the Configuration* and its relationship to the *Doubts* are discussed in the first two sections of our introduction.

The physical workings of the heavens were a matter of interest to Hellenistic scientists as well, and the most important Hellenistic astronomer, Ptolemy, devoted a special treatise to this subject, the *Planetary Hypotheses*. Here again, some hurried remarks on the part of one esteemed scholar have led to the erroneous view that Ibn al-Haytham's *On the Configuration* is but a reworking of Ptolemy's text. This opinion is refuted in the third section of our introduction.

We maintain that the *Configuration* is a text representative of a particular tradition within Islamic astronomy, the *hay'ah* tradition. This tradition and some of the people who contributed to it are discussed briefly in section four. Finally we note the medieval translations (section five) and explain our edition (section six).

1. Summary of *On the Configuration of the World*

It is clear from repeated references in chapter one that Ibn al-Haytham considers the principal deficiency of previous works on astronomy to be their failure to account physically for the workings of the heavens. The authors of these works have not totally neglected the physical structure of the universe. Quite the contrary: they speak of certain features of the heavenly bodies which Ibn al-Haytham will ignore in his treatise, such as "their distances from each other [and] the magnitude of their bodies" [passage 1]. These are important features of the physical universe, but limiting an analysis to these topics, in Ibn al-Haytham's opinion, begs the essential question of the correspondence of the motions and circles (spoken of by astronomers) to the actual physical components of the universe. Previous writers do indicate (*ashâra*) that these motions take place "on the surfaces of corporeal spheres" [passage 2] , but this is as far as they go. Ibn al-Haytham is willing to concede that previous astronomers were aware of a problem, but they never clearly addressed themselves to it. The problem, as it is articulated by Ibn al-Haytham, is to investigate each motion spoken of by Ptolemy (in the *Almagest*) in such a way that "that motion may appear to be the result of a spherical body that is moving with a simple, continuous, and unceasing motion" [passage 9]. These bodies may then be joined together to form a complete superlunar universe, which moves eternally without change or impediment.

It is of central importance for us to understand how Ibn al-Haytham himself defines the goals of his treatise, particularly in its relationship to previous works. Ptolemy, in his mathematical astronomy, has made free or abstract (*mujarradatan*) use of imaginary circles and points, in the sense that he has not bothered to relate these devices to solid spheres [passage 6]. Note that in the introductory chapter and, indeed, throughout the treatise, there is no criticism of these devices *per se*; we shall have more to say concerning this in the following section. Now certain of Ibn al-Haytham's predecessors have included some of the physical aspects of the heavens in their discussions, but their discourse on this matter is neither thorough nor transparent [passage 4]; it is merely affirmed (*mursal*) and, in addition, quite general (*jalîl*) [passage 1]. Thus these presentations are at best persuasive (*muqni* ᶜ) [passage 4], which, in medieval terms, implies the serious criticism that they fall short of demonstrative proof (*burhân*).

So much for the shortcomings of earlier writers. But what goals has Ibn al-Haytham set for himself? Basing himself on Ptolemy's geometrical theory, Ibn al-Haytham will attempt to present the planetary orbits in terms of the motions of spherical bodies which move with a simple, continuous, and unceasing motion [passage 9]. However, Ibn al-Haytham is *not* going to elaborate a comprehensive system, worked out to the last detail. The various motions can indeed, we are told, be joined together in such a way that the rules of cosmology will not be violated—that is, there shall be no "hindrance, repelling, or impediment" [passage 10]. Nonetheless, Ibn al-Haytham will present only "[general] remarks (*jumal*) . . . not the particulars of these" [passage 12]. It is assumed—perhaps one might say, it is hoped—that when these details are worked out, all will be in agreement [passage 12]. But this is not proven.

There are several important, acknowledged similarities between Ibn al-Haytham's plan, as outlined in the first chapter, and the works of his predecessors. Both undertake to present astronomy to a reader who is not himself interested or able to undertake research [passages 3, 7]. Ibn al-Haytham characterizes his own product as *jumal*, [general] remarks, i.e. a non-exhaustive descriptive account [passage 12]. This is, in fact, the same word which he has used, in passage 3, to describe the efforts of his predecessors. These characteristics of style—a general theory presented to a non-specialist—along with the particular emphasis given to cosmology, that is to say, the three-dimensional, real structure of the heavens, are, in fact, the chief distinguishing marks of the *hay'ah* tradition, that particular genre of Islamic astronomical literature to which our treatise may be said to belong.

The twelfth-century Andalusian philosopher Ibn Rushd (Averroes) must have had *On the Configuration* in mind when, at the beginning of his unpublished astronomical compendium that survives only in Hebrew translation [Steinschneider (1881), p. 728, and (1893), pp. 546-549; I consulted Ms. Oxford-Bodley Opp. Add. fol. 17 [Neubauer 2011], f. 104a], he included Ibn al-Haytham among "the authors of summaries" (*ba ᶜalei ha-qiṣṣurim*): "We have seen fit to compile here that [part] of the science of astronomy which appears to be requisite for [the attainment of] human perfection. This measure [of astronomy] is that measure which the authors of summaries in this science, such as Ibn al-Haytham and others, have written. However, what they wrote is devoid of verification nor is it [presented] according to the demonstrative method, for they intended only a [general] conception (*ṣiyyur*)..."

On the Configuration of the World is divided into two parts. The first part (chapters 2-8) deals with a variety of issues. Some, for example, the salient characteristics of the sublunar world as distinguished from the celestial regions, or the all-important definition of *falak* (orb), are clearly in line with the emphasis on cosmology expressed in the introduction (chapter 1). However, most of the material discussed in these chapters has no direct bearing on the question of the structure of the universe; rather, it encompasses the subject matter of astronomical texts—both introductory material (e.g. trigonometry) and ancillary material (e.g. mathematical geography)—presented in a clear and non-technical fashion. The second part (chapters 9-15) focuses on the planets and the fixed stars; here Ibn al-Haytham addresses more directly the problems that he has outlined in the introduction. But here too a lot of non-cosmological material is included. For example, the chapter on the moon discusses eclipses, including some remarks on the theory of vision, and the lunar phases. What we have before us, then, is no monograph reconciling physics with astronomy, but, rather, a fairly comprehensive non-technical introduction to the science of astronomy, which, however, does place special emphasis on "celestial physics."

Nevertheless, the stated objective of this treatise is to relate the observed orbits of the stars to the system of spheres and the related principles of motion accepted by the natural philosophers. How is this done? The same basic procedure is followed in all the chapters of part two. We shall analyse this in depth a little later on. Here we shall give only a brief synopsis.

Each planet has its own system of orbs. Ibn al-Haytham first introduces the orbs moving inwards from the outermost orb of the planet in question. These orbs are three-dimensional. The orbs are then reintroduced as the intersections of the first set of orbs with the planes of the circles described by the motions of the planet or the motion of the center of the epicycle. Thus the two-dimensional, "mathematical" components are matched with the three-dimensional, "physical" components. The motions of the orbs are then discussed, and the resultant motions of the planet are summarized.

In this whole analysis, Ibn al-Haytham is aided by the ambiguity of his definition of the technical term *falak* (orb) [Langermann, 1982]. Four definitions of this term are presented at the beginning of chapter three [passage 39], two of which are relevant to three dimensions, and two to two dimensions. There we are told that "orb" refers to any "round magnitude" (*miqdâr mustadîr*) of (1) a spherical body (*jism kuriyy*), (2) a spherical surface (*saṭḥ kuriyy*), (3) a plane of a circle (*saṭḥ dâ'irah*), or (4) a circumference of a circle (*muḥîṭ dâ'irah*). This key passage was cited by Hartner [1965, p. 762], who, however, mistranslated a key word in the following phrase (he rendered *bi-jumlatihi*, "in particular" rather than "in its entirety") and thereby somewhat misconstrued Ibn al-Haytham's intent. It would be appropriate at this juncture to illustrate what we have been saying by following in detail Ibn al-Haytham's "double presentations" through one of the planetary systems; let us take as our example the system of the moon.

The first presentation of the orbs of the moon [chapter ten, passages 210-220] utilizes the first two, that is the three-dimensional definitions. Five orbs are distinguished. The orb of the moon (1) occupies the space between the orb of Mercury and the sphere of fire. The parecliptic (2) is in the upper surface of this same orb, i.e. that touching the orb of Mercury. This orb is then divided into concentric sections, one of which (the lower) is surrounded by the other (the upper). The lower one is called the inclined orb (3). Within this lower orb, another orb is distinguished, again bounded by two concentric surfaces (though their center is different from the center of the inclined orb). The lower of these surfaces is the eccentric (4). Finally, within the eccentric a sphere is distinguished whose diameter is equal to the width of the eccentric "shell". This is the epicycle (5), which contains the moon itself.

These same orbs of the moon are presented anew in passages 221-226. In the previous discussion these orbs were distinguished by moving inwards from the outermost orb, and their presentation was in the context of spheres and spherical shells. Now these orbs are introduced by moving outwards from the eccentric, utilizing the intersections of the eccentric circle with the outlying orbs.

We are told first that there is another way of identifying the eccentric orb, namely that the imaginary circle traced by the epicyclic center about the eccentric's center is also called the eccentric orb [end of passage 221]. This is in accordance with the fourth meaning of *falak* given above. We then have the plane of this circle (the third definition of *falak*) intersect the "world" and take note of the circles that are produced in the other orbs.

A circle is produced in the epicycle, but we are really interested in the circles "parallel" (that is, homocentric) to this, i.e., another great circle on the epicyclic sphere, namely the one on which the center of the moon is to be found. (This will not necessarily be in the plane of the eccentric). This circle (and the word *dâ'irah*, or circle, is explicitly used) is also called the epicyclic orb. Similarly, two circles will be produced in the shell (the text reads *saṭḥ*, literally "surface") of the inclined orb and one in the first orb. This last circle (again specified in the text as *dâ'irah*) is also called the inclined orb.

It is important to note that the text uses both of the following expressions: "this sphere is called the orb" [end of passage 217] and "this circle is also called the orb" [passage 226]. The technical term *falak* (orb) is used both for the geometrical figures and actual bodies that are found in the heavens.

Although Ibn al-Haytham has succeeded in relating the mathematical constructs of the astronomers to the physical universe conceived by the philosophers, he has not attempted to present a system that meets the full requirements of medieval physics. For example, one of the basic rules of this physics can be stated as follows: to each body, only one motion can be assigned. But, to take an extreme case, four motions are associated in our text with Mercury's epicycle: the motion of the epicyclic diameter in alignment with the equant, the motions of the two orthogonal diameters that account in part for the motion of the planet in latitude, and the rotation of the epicycle itself.

One can thus characterize the accomplishment of our text in the realm of cosmology as the pairing of the circles that are found in the technical works on astronomy to their three-dimensional partners. No attempt is made in our text to resolve, or even to acknowledge, any of the manifest difficulties of medieval cosmology.

Before concluding our characterization of the text, we must say something about the Appendix (*ta ᶜlīq*). M. Schramm [1965, pp. 67–69, 138–41] was particularly interested in this short text, the Arabic text of which he published [p. 69]. We shall limit ourselves to a few brief remarks. First of all, we note that the Appendix is found in only one of the Arabic manuscripts and in none of the translations. Thus, though the Appendix is of interest in and of itself, it was of no influence to European readers of our text, and its influence on the Arabic audience is also questionable.

The Appendix, unlike the main body of the treatise, deals exclusively with the physics of the motions of the heavenly bodies. Short though it is, it addresses an issue not covered in our text, namely the interaction of the moving bodies in the heavens. The concise remarks of the Appendix do not contain any criticism of Ptolemy.

It is extremely difficult to determine on the basis of internal evidence (which is all that we have) the exact relationship of the Appendix to the treatise. On the one hand, Ibn al-Haytham does imply at the end of the first chapter that the goal of *On the Configuration of the World* is a set of principles. Perhaps the Appendix contains the statement of these principles. Yet one would expect principles of this sort to be brought up in the treatise itself, not saved for an appendix. Some indeed are found in the text: the impossibility of the vacuum, for example, is stated in chapter two [end of passage 14]. Our feeling at the moment is that the Appendix represents an effort at a concise formulation of physical principles, which was written independently of our text.

2. *Al-Shukûk ᶜalâ Baṭlamyûs*

We can best demonstrate the irrelevance of our text to the outstanding problems of medieval cosmology by indicating several points of disagreement between our text and another work of Ibn al-Haytham, *Doubts concerning Ptolemy* (*al-Shukûk ᶜalâ Baṭlamyûs*; Arabic text published by A. I. Sabra and N. Shehaby; see also Pines, 1964). In this work, as the title suggests, many criticisms of Ptolemyare brought forth, some of which strike at basic cosmological issues. Many of the sharp criticisms of Ptolemy which are developed in the *Doubts* can, in fact, be directed equally well at *On the Configuration*, which faithfully mirrors the astronomical theory of the *Almagest* .

One fundamental criticism raised in the *Doubts* concerns Ptolemy's assignment of motions to points, lines, and planes, all of which are mathematical constructs. Ibn al-Haytham contends in the *Doubts* that the motions found in the universe, and the phenomena in the heavens that result from those motions, can come about only from the motions of real, i.e. sensible, objects. Thus, for instance, in criticizing the prosneusis of the moon, Ibn al-Haytham remarks [p. 16]:

> The diameter of the epicyclic orb is an imaginary line. An imaginary line does not of its own (*bi-dhâtihi*) move with a sensible motion which produces something which exists (*maᶜnan mawjûdan*) in the world. Similarly, the plane of the epicyclic orb is an imaginary plane, and an imaginary plane does not move with a sensible motion. Nothing moves with a sensible motion to produce something which exists in the world other than a body which [itself already] exists in the world.

We note that in our text the prosneusis is attributed to the moon without criticism [chapter ten, passage 238]. Nor is this the only motion to be assigned to an imaginary line in our text. Note, for example, the motions of each of the orthogonal diameters of the epicycles of Mercury and Venus, which contribute to the motion of the planet in latitude [chapter eleven, passages 310-312; chapter twelve, passage 333].

Similarly, the equant comes under fire in the *Doubts*. For the supposition that the equal motion of the epicyclic centers of the five planets (the sun and the moon are excluded) takes place about a point other than the deferent center implies that its motion on the deferent circle is nonuniform. Ibn al-Haytham concludes [p. 25]:

> This is contradictory to what he [Ptolemy] established concerning their motions, because he established that all their motions are uniform.

In our own text, the equant is included without criticism in the models of each of the five planets [chapter eleven, passage 295; chapter twelve, passage 328; chapter thirteen, passage 348]. Moreover, it is to my mind quite clear that, in *On the Configuration*, the equant has physical significance assigned to it, and the vector (to use a modern term; the Arabic *khaṭṭ* simply means "line") issuing from it is imagined to function mechanically in the planet's motion. Note, for example, that in the model for Venus [passage 328], the equant (*markaz mu ᶜaddil al-masîr*, literally "the center of equal motion") is also called the dirigent* center; the Arabic *mudîr* is a transitive form. The vector extending from the equant to the center of the epicycle is also called the dirigent (*mudîr*) line. The mechanical implication of this choice of terminology is quite transparent. Further evidence for our interpretation is found in passage 358, where Ibn al-Haytham lists the motions of the superior planets. The second motion, that of the deferent (*al-ḥâmil*), is called *al-mudîr*, that is, that which turns, the epicyclic orb. Here *mudîr* is used to describe the indisputably mechanical function of the deferent.

Again, we note that the oscillation of the plane of the inclined orb, an essential component of Ptolemy's latitude theory for Mercury and Venus, is strongly attacked in the *Doubts* [pp. 36-7]:

> This is utter nonsense and contradicts his previous doctrine that the heavenly motions are equal, continuous, and unceasing. For it is impossible that anything but a body move with this motion, for only existing [that is, real] bodies can have sensible motions. These two contrary motions [the oscillation of the plane in two opposite directions] can be either natural or voluntary (*'ikhtiyariyyah*), because the motion of every

* I am indebted to my colleague F. J. Ragep for this term. The Arabic word is *al-mudîr* .

> body which moves by itself must be either natural or voluntary. In the world of generation and corruption there is no body which moves with two contrary natural motions. It is a forbidden impossibility, even more forbidden for the heavenly bodies, for these are more remote from contrariety. If these two motions are voluntary, it follows that the body which moves with them be made of two or more contrary substances. This is impossible for the heavenly bodies.

Yet the oscillation of the plane of the inclined orbs of Mercury and Venus is described without criticism in our text [chapter eleven, passage 308; chapter twelve, passage 335]. These and other criticisms lead Ibn al-Haytham, in the *Doubts*, to proclaim that the configuration of the heavens which is presented in the *Almagest* is simply a "false configuration" (*hay'ah bâṭilah*) [p. 34]. In contrast, in our text, no criticism at all of Ptolemaic cosmology is to be found. To the contrary: in the final chapter, when summarizing all of the motions which exist in the universe, Ibn al-Haytham reminds us that he has relied upon the investigations and observations of Ptolemy for all of the heavenly motions.

Indeed, no astronomer or philosopher other than Ptolemy is mentioned by name in *On the Configuration*. It is also interesting to note that there are only two places in our treatise where more than one opinion is presented concerning one of the issues at hand. The first occurs in chapter four [passage 101], where a non-Ptolemaic determination of the of the boundaries of the seven climates is given side by side with Ptolemy's. This clearly stems from the controversy among Islamic scientists concerning the possibility of human civilization south of the equator. The other occurrence is in chapter nine [passage 198], where Ibn al-Haytham cites the views of the later astronomers who disagree with Ptolemy and claim that the sun's apogee is not fixed. It may be suggested that in these two cases Ibn al-Haytham recognized that substantial evidence existed in favor of the non-Ptolemaic viewpoint, in the first case accumulated by Arab voyagers, and in the second place resulting from new observations. (See our footnotes to these two passages in the English translation.) Yet even in these instances, Ptolemy's opinion is presented, and Ibn al-Haytham does not say which opinion he holds to be the correct one.

3. *The Planetary Hypotheses*

It is a well-accepted fact that the astronomical work of Ptolemy can be subjected to the following division: a purely mathematical scheme for the representation of the heavenly motions and an attempt at a physical description of how the heavens operate. The first topic is the concern of the *Almagest*, and the second is taken up in the *Planetary Hypotheses*. Like all generalizations, the one under discussion must not be taken absolutely. Thus some physical principles are invoked in the *Almagest*, especially in Book I [trans. Toomer, pp. 38-47]. Many observations and parameters belonging to the mathematical theory are to be found in the *Hypotheses*. On the whole, however, we may accept this bipartite scheme as valid.

Given this situation, the question of the influence of the *Hypotheses* on our text is a natural one. We have already pointed out that, from the point of view of its overall contents, our text is not devoted primarily to the physical description of the heavens. It is concerned with presenting the science of astronomy in all of its non-technical aspects, and the physical structure of the heavens, despite the special emphasis which it enjoys, is just one of a number of topics to be treated.

This notwithstanding, one should still be interested in making some comparison of our text with the *Hypotheses*, since that text is, after all, Ptolemy's own and rather detailed account of the physical side of astronomical theory. In any event, such a comparison is mandated by a misconception of our text which was introduced into the scholarly literature some seventy-five years ago by Pierre Duhem and which has been repeated ever since by almost all those who have had occasion to refer to our text.

Duhem devotes some ten pages of the second volume of his *Le Système du monde* [1914, pp. 119-129] to our text, which he justifiably entitles *Résumé d'Astronomie*. Duhem relied on M. Steinschneider's descriptions and the his own study of parts of the Vatican Latin manuscript (the translation, from the Hebrew, by Abraham de Balmes; see below, section 5). Duhem's interest is confined to the second part of our text. He translates some passages and summarizes others.

Duhem's conclusion concerning our text is as straightforward and unambiguous as it is wrong [p. 124]:

> Ces quelques extraits du *Résumé d'Astronomie* nous renseignent pleinement sur l'origine et la nature des

> mécanismes que décrit ce traité. Ces mécanismes sont empruntés de toutes pièces aux *Hypothèses des astres errants* de Ptolomée.

Thus Duhem is led to an even more general statement [p. 125]:

> En ce cas, donc, comme en tantes d'autres, la Science arabe n'a fait que réproduire des enseignements qu'elle avait reçue de la science grecque.

The very eminent orientalist C. A. Nallino is also guilty of the misconception that *On the Configuration* derives from the *Hypotheses*. In an excellent article which he prepared for Hasting's *Encyclopaedia of Religion and Ethics* [vol. 12, 1921—which contains, I must add, many important leads for the history of Islamic astronomy and astrology which remain to be fully exploited—he states [p. 99]:

> The Aristotelian conception of solid spheres was introduced for the first time into a purely astronomical treatise by Ibn al-Haitham; and he, in his unedited compendium of astronomy, gives the definition which was accepted afterwards by all the other writers of elementary treatises: 'A celestial sphere (*falak*, plur. *aflâk*) is a body completely spherical, bounded by two parallel spherical superficies having the same centre.' In this, as he himself says, he drew his inspiration from the *Hypotheses Planetarum* of Ptolemy; in fact, as we see from a passage from Naṣîr ad-dîn aṭ-Ṭûsî, he followed the *Hypotheses* also in expounding how the celestial motions can be represented, and also by supposing simple equatorial zones of those complete spheres, so that the spheres of the epicycles become, as it were, tambourines (*duff*) rotating on their own axes, and the other spheres like armils. This second form of representation was soon abandoned, as contrary to the principles of natural philosophy.

The short quotation from passage 39 and the general description of the treatise allow little doubt that *On the Configuration* is intended. In a footnote Nallino cites "the preface itself of Ibn al-Haitham" and "the attestations of Muḥammad al-Akfânî al-Saḥâwî, *Irshâd al-qâṣid*, p. 85." I do not know which manuscript of our treatise was seen by Nallino. However, the *Hypotheses* is never mentioned in our text. Ptolemy's "book on mathematics" (*kitâb fî al-ta ᶜlîm*), which is cited several times, including in the introduction [passage 6], undoubtedly refers to the *Almagest* [Sezgin (1978), p. 88]. Note further that Ibn al-Haytham says [passage 9] that Ptolemy "avoided the use of any bodies." How could such a remark be made with regard to the *Hypotheses*? Moreover, neither the "tambourines" nor any of the other shapes that are suggested in the *Hypotheses* are discussed or even noted in *On the Configuration*. I was able to check a later edition of the work of al-Akfânî (or Ibn al-Akfânî, d. 1348; ed. ᶜAbd al-Laṭîf Muḥammad al-ᶜAbd, Cairo, 1978, pp. 142-43), where I found only the following statement: "The ancients were always constricting themselves to abstract circles for the configuration of the orbs, until Ibn al-Haytham openly declared (*ṣarraḥa*) their [the orbs'] corporeality. He recorded their necessary characteristics and their circumstances, and the later [scholars] have followed him with regard to this." Nallino provides us with no source reference for the passage quoted from al-Ṭûsî. We do know, however, that al-Ṭûsî had available to him other works of Ibn al-Haytham in which the *Hypotheses* does figure [Sabra (1979), p. 390]. Such inaccuracy is quite out of character for Nallino, and I cannot refrain myself from speculating that he has been misled by some later sources.

We may also mention a more recent work which discusses Ibn al-Haytham's achievements in celestial physics, namely Henri Corbin's *Histoire de la philosophie islamique* [1986, pp. 214-217]. *On the Configuration* is not cited by name, but it appears to be the source for Corbin's remarks concerning Ibn al-Haytham's alleged accomplishment. Here again the *Planetary Hypotheses* is mentioned [p. 215]:

> L'extrême intéret de la situation, c'est que d'une part le problème se posait à Ibn al-Haytham en termes de physique céleste (comme physique essentiellement qualitative), à la suite du Ptolomée des *Hypotheses Planetarum*, lequel, lui aussi, recourait à une physique céleste déduite de la nature de la substance qui forme le ciel, et ne faisait que substituer sa propre physique à celle du *De Cælo* d'Aristote.

We need not go any further into Corbin's analysis, other than to point out that the term *falak al-aflâk*, which Corbin wishes to connect to Ibn al-Haytham, is nowhere to be found in *On the Configuration*.

It thus emerges that several scholars have connected *On the Configuration of the World* with Ptolemy's *Planetary Hypotheses*. In particular, it seems that Duhem's view has strongly influenced the portrayal of *On the Configuration* among students of medieval cosmology; the widely-cited survey article of E. Grant [1978, p. 281] may serve as just one example. Specialists in the Arabic literature also repeat this misconception, for example, F. Sezgin [1978, p. 95]. It is our contention that Ibn al-Haytham's work represents an independent effort, not modeled after, and, possibly, not even influenced by the *Hypotheses*. Because the belief that *On the Configuration* derives from the *Hypotheses* is so widespread, we devote here a separate section of our introduction to a survey of the *Hypotheses*, in order to demonstrate just how different the two works are in form and content.

However, since we have just had occasion to list a number of scholars who have expressed what are, in our opinion, inaccurate appreciations of *On the Configuration*, it seems to be appropriate at this juncture to single out the careful and perspicacious remarks of two other scholars. Matthias Schramm [1963, pp. 63-69] studied the Arabic original of our text; his brief remarks are quite accurate and valuable. Also worth citing is this context is a short study published by S. Pines [1946] which, because it appeared in Hebrew, is not as well-known as some of the other publications of that greatly lamented scholar. In that article Pines does not address our text as such, but he does refute Duhem's [1914, p. 130] obviously prejudiced contention that Ibn al-Haytham, like the other "astronomes arabes", was a "realist" in so far as he wished the bodies spoken of in Ptolemy's *Hypotheses* to correspond with the actual state of affairs in the heavens. Pines quotes a passage in al-Bayhaqī's *Ṣiwân al-ḥikmah* (ed. M. Shafī^{c}, Lahore, 1935, I, p. 78) to the effect that Ibn al-Haytham, in his very last composition, expressed his openness to any model which may adequately account for the observed stellar motions, demonstrating, according to Pines [p. 103], that "in contradiction to the opinion of that one [Duhem], the Arab scholar [Ibn al-Haytham] absorbed not only the body of the doctrines of his teachers [i.e. Ptolemy and the other Greeks] but also their methodological carefulness, their fastidiousness in fixing boundaries for the scientific imagination. It is not impossible that [we are confronted] here not only with drawing upon [earlier] sources, but also with individual thought..." (Since the issue has now been raised, we may add that we fully concur with the reservations expressed by N. Jardine [1982, p. 172] concerning the relevance of the realist-instrumentalist dichotomy to the investigation of medieval cosmology.)

We shall here go through the text of the *Hypotheses*, describing the contents of its various sections. We shall pause for a closer look at a few of the more crucial passages, those which contain material which, we believe, forms the core of the approach of the *Hypotheses* and, therefore, we have a right to expect similar material in any work deriving from the *Hypotheses*. By no means are we offering a complete analysis of the *Hypotheses*; such an undertaking surely warrants an independent scholarly effort. We hope nonetheless that our discussion will form a useful supplement to the rather meager amount of modern analysis of the *Hypotheses*, which has been recently summarized by N. Swerdlow [1977, p. 115, n. 12].

The translations which I offer here have all been made from the Arabic text, which has been published with variants (but not edited) by B. R. Goldstein [1967]. References are given to page and line in Goldstein's version, and also to J. Heiberg's text [Nix,1907]. Key words such as *falak* (orb) and *kurah* (sphere) are translated exactly as they are in *On the Configuration*, and, indeed, as they are in the selections that we quote from Arabic texts elsewhere in our introduction.

In the opening section of the *Hypotheses* [Nix 73:1-75:6; Goldstein 13:1-14:9] Ptolemy explains that his goal in this work is to explain mechanically the workings of the whole universe and each of its parts. Such an elaboration will be useful both to those who wish to picture these things for themselves and for those who wish to construct an instrument that will represent the heavenly motions. Ptolemy proposes to begin his project with a discussion of the "universal motion" (*al-ḥarakah al-kulliyyah*), for it encompasses all of the particular motions. This will then serve as an example for many of the issues to be raised presently.

In the subsequent paragraph [Nix 75:7-77:21; Goldstein 14:10-15:9] Ptolemy introduces the orb of the celestial equator (*falak muᶜaddil al-nahâr*), defined as one of the great circles drawn on the center of the world. Within this orb he distinguishes the ecliptic orb, whose inclination is given as 23°51'20". Ptolemy also points out the distinction between a single revolution of the whole world and our earthly nycthemeron. The latter is the basis of astronomical parameters, and its difference from the former is due to the motion of the sun.

We wish to note here some points of difference between our text and the Arabic *Hypotheses* with regard to terminology. It bears repeating that one of the more important tools available to historians of early science are the technical vocabularies, by means of which it is possible to trace the influence of key texts and/or ideas, when these same terms are found in works of a later generation, and also to detect changes in outlook, when the old vocabulary takes on new meanings. The *Hypotheses* calls the 360 divisions of the circle *azmân*, that is, units of time. In our text, these units are called either *juz'* (part) or, more rarely, *darajah* (degree). One important phrase employed in the *Hypotheses*, *makhṭûṭ ᶜalâ markazihi* (drawn about its center), with its connotation of a two-dimensional geometrical representation, is not found in *On the Configuration.*

Ptolemy now turns to the planets and the fixed stars [Nix 77:22-81:2; Goldstein 15:10-16:18]. He speaks first of their simple (*basîṭ*) motions, giving the numerical data for their anamolistic returns in both solar and Egyptian years. We remind the reader that no such parameters are to be found in our text.

Ptolemy then proceeds to a fuller analysis, treating each one of the seven planets separately [Nix 81:3-107:8; Goldstein 16:19-26:3]. He generally introduces the discussion by mentioning the sphere of the planet ("the sphere of the sun," etc.), but the subsequent discussion concerns orbs and circles. He also provides various parameters, such as eccentricities and epoch values. These parameters have all been thoroughly analysed by Neugebauer [1975, vol. 2, pp. 900-922].

If any portion of the *Hypotheses* could have served as a model for our text, it is this portion. Despite the obvious differences, most notably the absence of all mathematical data from our text, the general presentation in terms of imagining certain circles to be in certain orbs may perhaps be seen to bear some superficial resemblance to our text. Clearly, it was this section of the *Hypotheses* that Duhem had in mind when he put forth his claim. Nevertheless, we feel that if one were to compare the discussions in this section of the *Hypotheses* with the corresponding material from Ibn al-Haytham's text, the differences in content and in style would become apparent. To this end we translate here the passage from the *Hypotheses* which deals with Venus [Nix 90:8-94:6; Goldstein 20:15-21:24] and invite the reader to compare it with the beginning of chapter twelve of our text. These two sections are both relatively uncomplicated and short, yet adequately representative of their respective texts.

> Regarding the star Venus, let us also imagine that it has an orb whose center is the center of the ecliptic orb, moving in its plane and about its center with an equal

motion from west to east, similar to the motion of the fixed stars. Let this orb set in motion, by means of its own motion, another orb inclined to it and about its center but not leaving it (*ghayr zâ'ilan ᶜanhu*). Let the inclination of its plane comprise an angle of six parts of such measure that the right angle has 90 parts. Let there be in the plane of the inclined orb a diameter [extending] from the northern limit (*muntahâ*) to the southern limit, on which we discern two points between the center of the ecliptic orb and the northern limit. Let the line between these two points be equal to the line [drawn] between the center of the ecliptic orb and that one of the two points which is close to it. Let the eccentric orb be drawn (*makhṭûṭ*) on that one of the two points which is closer to the earth, not leaving [its position] and not moving. Let the ratio of its semi-diameter to the line between its center and the center of the ecliptic orb be like the ratio of 60 to 1. Let the epicyclic orb move about that one of the two points which is further from the earth with an equal motion, and let the position of its center always be on the eccentric orb according to the manner in which the diameter which we mentioned keeps to the ecliptic orb [the text here is quite problematical; cf. Goldstein 20:27] with the rate of the excess of the sun's motion over the motion of the fixed stars, in equal times. Let us also imagine a small circle in the sphere of the epicyclic orb, on its center and in the plane of the inclined orb. Let the line which passes through its center and that one of the two points which we mentioned that is further from the earth, on which this orb moves with an equal motion, pass through this small orb at exactly two points, which are those that are called the apogee and closest distance. Let us also imagine another small orb whose center is the center of this orb. It moves on its surface with an equal motion from the apogee in the direction in which the world moves, with a motion equal to the motion of the epicyclic orb which we mentioned. This orb sets in motion by means of its own motion another orb inclined to it, whose center is its center, and it does not leave this orb. Let the inclination of this orb comprise an angle of $3\frac{1}{2}$ parts of such measure that the right angle has 90 parts. Let the ratio of the semi-diameter of the eccentric orb to the semi-diameter of the epicyclic orb be like the

> ratio of 60 to $43\frac{1}{6}$. Let the star move about the center of this orb with an equal motion from the apogee in the direction opposite of that in which the world moves, with a motion equal to the combined motions of the epicyclic orb and the star. In 35 Egyptian years and 33 nycthemerons it made 57 returns. This is an approximation, for it exceeded what we have said, when one inquires, by one minute. The distance of the apogee of the place of eccentricity (*mawḍi ᶜ al-khurûj ᶜan al-markaz*) from the point of the vernal equinox, as it keeps to the ecliptic orb [i.e. measuring from west to east, in the order of the signs] in the first year after the death of Alexander, in the first of the Coptic month of Toth, at noontime in Alexandria, was 50°24'. So also was the distance of the northern limit from this point. The distance of the center of the epicyclic orb from the apogee of the place of eccentricity as it keeps to the ecliptic orb, was 177 parts and 16 minutes. The distance of the northern limit of the small inclined orb from the apogee of the epicyclic orb, against the order of the signs, was 87 parts and 16 minutes. The distance of the planet from the northern limit on [emending *'ilâ* to *ᶜalâ*] the small orb, as it keeps to the ecliptic orb, was 168 parts and 35 minutes.

Let us briefly note some of the salient differences between this section of the *Hypotheses* and the corresponding part of our treatise. Conspicuously absent in the *Hypotheses* is the "double presentation" of Ibn al-Haytham, namely the introduction of the orbs first in their three-dimensional and then again in their two-dimensional aspects. Also lacking is the succinct correlation of the motions of the planet to the various orbs. On the other hand, numerical values for the eccentricites and the rates of motion are given by Ptolemy but not by Ibn al-Haytham.

We now come to what Goldstein has named part II of Book I, the famous section of the text discovered by that scholar and missing from Nix's edition. This section [Goldstein 26:4-35:14], which has been translated and analysed by Goldstein [pp. 4-12], deals primarily with the distances, sizes, and *arci visionis* of the planets, none of which appears in our text. The discussion in the "Aristotelian introduction" (as Goldstein [p. 4] calls the introductory passages of part II) agrees with our text on basic issues, such as the unchanging nature of the heavenly motions. This is to be expected. For the most part, however, the material found in this section also is missing from *On the Configuration.*

In Book II Ptolemy undertakes to establish the shapes of the bodies that carry out the heavenly motions. He sets forth several basic principles, for example, the principle that ethereal bodies do not accept being acted upon (*infi ᶜâl*), nor do they change. He lays down the fundamental law that is echoed in chapter one of our text, and in the appendix as well, although in practice it is violated in the body of our text; this rule also figures prominently in the *Doubts*. It states [Nix 112:12-16; Goldstein 36:19-20]:

> For each of these motions, which are different in quantity or kind, there is a body that moves freely on poles and in space and which has a special place . . .

Ptolemy then postulates two possible paths of approach to the physical explanation of the workings of the cosmos. This duality is referred to several times in the ensuing discussion. Ptolemy writes [Nix 113:16-33; Goldstein 37:9-17]:

> The first of them is to assign a whole (*tâmm*) sphere to each motion, either hollow like the spheres that surround each other or the earth, or solid and not hollow like those which do not contain anything other than the thing [itself], namely those that set the stars in motion and are called epicyclic orbs. The other way is that we set aside for each one of the motions not a whole sphere but only a section (*qiṭ ᶜah*) of a sphere. This section lies on the two sides of the largest circle which is in that sphere, namely that from which the motion is longitude [is taken]. That which this section encloses from the two sides is [equal to] the amount of latitude. Thus the shape (*shakl*) of this section, when [taken] from an epicyclic orb, is similar to a tambourine (*duff*). When taken from the hollow sphere, it is similar to a belt (*niṭâq*), an armband (*siwâr*), or a whorl (*fulkah*), as Plato said. Mathematical investigation shows that there is no difference between these two ways that we have described.

The section which follows these remarks is probably the most troublesome part of the *Hypotheses*. The first topic is the nature and operation of the poles of the various spheres. Aristotle's views are quoted and discussed. This issue is very important for an understanding of the way in which the motions of the various spheres are transmitted to each other and interact. No discussion at all of this problem is found in our text. It was of concern, as we shall see, to at least one of Ibn al-Haytham's predecessors, Thâbit bin Qurra, who, interestingly enough, is the redactor of the Arabic version of the *Hypotheses*. The original translator is apparently unknown [Goldstein, p. 5; Sezgin (1978), p. 95].

Ptolemy also puts forth the theory that the planets possess souls. He then turns to the problem of how all these various spheres fit together into a workable system, and he seems to favor the free motion of individually ensouled planets. The famous example of the bird is brought forth in order to illustrate how motions can be transmitted to the outlying parts of a system without necessarily interfering with the interlying parts. Ptolemy summarizes his views in this key passage [Nix 120:6-22; Goldstein 41:2-8]:

> For this reason we must construe the notion (*'amr*) of the celestial animal (*al- ḥayawân al-falakiy*). We hold that each one of the stars in its [individual] ranking possesses an animate (*nafsâniy*) power. It moves by itself, bestows motion upon the bodies naturally connected to it beginning with that which is close to it, and passes it on to that which is adjacent to it. For example, it bestows motion first on the epicyclic orb, then on the eccentric orb, and then on the orb whose center is the center of the world. This motion which it bestows is different in many places. For the motion of reason in us is not exactly like the motion of impulse (*'inbi ᶜâth*), nor is this motion like the motion of the nerves, nor is this motion like the motion of the foot. Rather they have some difference in their outward disposition (*fî maylihâ 'ilâ khârij*).

The possibility is then raised of constructing an instrument that can demonstrate how these motions take place in space. Ptolemy again reminds his readers that his analogies are only an attempt to make the planetary models more easily understandable.

Finally, Ptolemy is ready to move on to a description of the planetary systems themselves. In order to avoid needless repetition, he first puts forth a generalized explanation of the types of orbs that he will be using [Nix 122:13-123:6; Goldstein 42:9-23]:

> These bodies which move from east to west on the poles of the orb of the celestial equator and of necessity carry all that they encompass in the direction of the motion of the whole, are called by a general name, that is, "the mover" (*al-muḥarrik*). The first of these bodies is that which sets in motion the sphere of the fixed stars. The second is that which sets in motion the eccentric sphere of Saturn. The third is is that which sets in motion the eccentric sphere of Jupiter, and so on successively. Each one of the bodies which lie under this body is named according to the accidents which befall each one of them, i.e. by comparing their position with the position of the ecliptic orb. That is, some of them which encompass the earth rotate on the axis of the ecliptic orb itself and are called "similarly arranged" (*al-mutashâbihat al-marâtib*). Some of them are centered on the center of this [ecliptic] orb but do not rotate on its axis. They are called "inclined orbs."
>
> Some of them [here I follow the Leyden manuscript] are not on its center and do not rotate on its axis. Of this [second group], some rotate on an axis parallel to the axis of the ecliptic orb and are called by a special name, "eccentric orb." Some of them rotate on an axis which is not parallel to the axis of the ecliptic orb. They are given a name different from the name of the first [group], that is, "contrarily arranged" (*ḍidd al-mutashâbihat al-marâtib*).
>
> Those which do not encompass the earth are given a general name, which is "epicyclic orb." Some of them move on an axis which is parallel to the inclined orb which we have mentioned and are called "not inclined." Some of them move on an axis which is not parallel and are called [following again the Leyden manuscript] "differently inclined" (*mukhtalifat al-mayl*). Those of them which encompass the luminous bodies are called "movers of the stars."

We wish to make the following observations concerning *On the Configuration*. Ibn al-Haytham never even alludes to the souls of the planets or the spheres. There is no mention of an instrument which may represent the motions of the orbs, nor any indication that the models that are presented are to be understood as approximate mechanical analogies. Finally, there is no attempt in our text at a classification like that found in the passage just quoted. Several of the orbs in Ptolemy's scheme are not found at all in our text, particularly the several types of epicycle; our text knows of only one type of epicycle. Also, some of the more important technical terms are different, especially *al-mutashâbihat al-marâtib*, which seems to fulfill the same role as the parecliptic (*al-mumaththal*) that appears in our text. The Arabic *Hypotheses* uses here the verb *dhahaba* to convey the meaning "to carry, to set in motion," whereas Ibn al-Haytham employs *harraka*, occasionally substituting *naqala* or, more rarely, *qâda*.

Ptolemy next elaborates upon the various sets of orbs, beginning with the model for Saturn. The identical set of devices will serve for Jupiter, Mars, and Venus. The models for the sun, Mercury, and the moon are developed separately. Note that in *On the Configuration*, Ibn al-Haytham, like Ptolemy in the *Hypotheses*, treats the three superior planets together in the same chapter; Venus, however, is given a chapter to herself. This not merely an editorial question regarding the arrangement of material, but a stark consequence of the fact that Venus does not possess the same physical setup as the superior planets, especially in all that concerns her motion in latitude. As a matter of fact, Venus' model may be considered to be more similar to that of the superior planets or to that of Mercury, depending upon one's chief area of concern. As far as the arrangement of circles used to represent the star's motion in longitude is concerned, Venus' model is identical to that of the superior planets. However, if one takes into consideration the rather complex configuration devised by Ptolemy in order to account for the motion in latitude—something, as it seems to me, which would be required of any comprehensive three-dimensional model—then the case of Venus is like that of Mercury. (In addition, the way in which Venus' model keeps pace with the sun is identical to that of Mercury.) An anonymous late medieval Hebrew astronomical treatise, preserved, to the best of my knowledge, uniquely in a manuscript at Budapest, Kaufmann A520*, reports (pp. 310-311) a general difference of opinion concerning this very subject: "The experts in astronomy disagree concerning the configuration (or: model; Hebrew *t ekhunah*) of Venus. Some say that it is identical to the configuration of Mercury, but others state that it is the same as the configuration of the other planets." It thus emerges that Ibn al-

* A film of this (no. 15170) is available at the Institute of Microfilmed Hebrew Manuscripts, Jerusalem.

Haytham is, at least on this point, more fastidious than Ptolemy (in Book II of the *Hypotheses*) concerning the physical representation of the planetary systems. Ibn al-Haytham describes in detail Venus' motion in latitude, and he notes that, in general (for Venus, unlike Mercury, does not have a moving deferent) the models for the two planets are the same [passage 337]. Nevertheless, as we have already seen, Ibn al-Haytham does not present in *On the Configuration* a complete and satisfactory physical account for the latitudinal motion. He does appear to have treated that subject more fully in his lost treatise *On the Movement of Iltifâf* [Sabra,1979].

It is also important to note that nearly all of the discussion in the *Hypotheses* is carried out with reference to diagrams. The orbs, circles, and points spoken of are referred to in the text by letters of the alphabet. Since only some of the diagrams are to be found in the surviving Arabic manuscripts, it may not be too bold to raise the question whether any medieval astronomer ever fully understood these discussions. Nix did attempt to reconstruct the diagrams, but his efforts have been justly criticized by Neugebauer.

The last matter which we shall describe in detail is the count of the total number of motions that are found in the celestial realm. Neither in the *Almagest* nor in the *Hypotheses* does Ptolemy supply a list of all the heavenly motions. There is a list at the end of *On the Configuration* [passages 380-381]; it represents Ibn al-Haytham's own effort and is based on the *Almagest* . Now in the *Doubts* [pp. 39-41] Ibn al-Haytham lists all of the heavenly motions that are found in both the *Almagest* and in the *Hypotheses*. In the former work the sum total comes to 36: one for the sun (anomaly); five for the moon (nodes; latitude; apogee; anomaly; prosneusis); five for each of the three superior planets (apogee; longitude; epicycle; anomaly; inclination of the epicycle); seven for Venus (five like the superior planets; inclined plane; slant of orthogonal epicyclic diameter); and eight for Mercury (seven like Venus; dirigent orb).

This account will fit in very well with that which appears in our text after we make the following notes. In our text, each one of the planets is assigned one more motion. This "extra" motion comes about from Ibn al-Haytham's insistence in our text that the proper motion of the eccentric and the motion which results from the eccentric's motion vis-à-vis the ecliptic be counted separately. This is made clear in all of the separate listings found in the chapters on the planets (e.g. for the moon, chapter ten, passage 246, motions 3 and 4; for the superior planets, chapter thirteen, passage 358, motions 2 and 3). Adding these seven motions brings the total (based on the *Almagest*) to 43. If we then add the motion of the highest orb, the motion of the orb of the fixed stars, and the two rectilinear, sublunary motions, we will have 47 motions, exactly as in our text [passage 380].

A little later in the *Doubts,* [pp. 42-43] Ibn al-Haytham produces a list of the planetary motions found in the *Hypotheses* that apparently derives from the summary of planetary theory found in that book [Nix 81:3-107:8; Goldstein 16:19-22:3]. Here Ibn al-Haytham finds only 26 motions. He accuses Ptolemy of omitting the following ten motions: (1) the moon's prosneusis; (2-7) the three components of the motion in latitude for both inferior planets, namely the motion of the inclined plane and the motions of the two orthogonal epicyclic diameters; (8-10) the motion of the epicyclic diameter for each of the superior planets.

It is thus quite clear that the list in our text derives from the *Almagest*. We may also note that in the *Doubts* both lists are rejected since neither offers the one to one correspondence of bodies to legitimate motions, which is one of the key cosmological demands stated in the *Doubts* .

Two enumerations are found at the end of the *Hypotheses*. However, these are both lists of the bodies found in the universe, not of motions, as in our text. In the first list [Nix 141:27-35; Goldstein 53:14-18] Ptolemy counts spheres (*'ukar*) and finds 41 of them: eight "movers" (*muḥarrikah*) and, additionally, one for the sun, one for the fixed stars, four for the moon, five each for Venus and the superior planets, and seven for Mercury. Nowhere in our treatise does Ibn al-Haytham count up spheres, whether for an individual planet or for the entire cosmos. This would be a strange omission if he were attempting a comprehensive analysis of the universe on the model of the *Hypotheses*.

In his second list [Nix 142:1-8; Goldstein 53:18-22] Ptolemy counts bodies (*'ajsâm*), of which he finds 29: three hollow spheres (*'ukar mujawwafah*), namely that which moves the fixed stars, the sphere of the fixed stars, and that containing the rest of the ether; and 26 "disks" (*manshûrât*; literally, "sawed off sections"), of which the sun has one, Mercury five, and the other planets four apiece. Once again, we note that in our text Ibn al-Haytham does not count bodies, nor do the important terms *kurah mujawwafah* and *manshûrât* ever occur in our treatise.

The *Hypotheses* concludes with a short reprise concerned with the physical form of the tambourines and armbands which have been proposed as illustrations of the types of bodies found in the heavens. Finally, there is a discussion of the lost tables that were apparently appended to the *Hypotheses*.

There remains one further matter to be considered. The Arabic version of the *Hypotheses* went by two names [Sezgin (1978), p. 94]. One of these is (*kitâb*) *al-manshûrât*. We have just had occasion to note that *manshûrât*, a term that clearly carries great cosmological import, is not to be found in *On the Configuration*. The term which figures in the other Arabic title, (*kitâb*) *al-iqtiṣâṣ*, does appear three times in *On the Configuration*, in passages 6, 321, and 359. However, *'iqtiṣâṣ* means simply "accurate presentation"; it carries no specific cosmological significance, and its usage, particularly in view of all of the evidence that we have presented in this section, offers no serious challenge to our conclusion.

A full and detailed study of the *Hypotheses* remains high on the agenda of historians of astronomy. Our survey has had the limited aim of dispelling the notion that *On the Configuration* is but a reworking of the *Hypotheses*. Now there are bound to be some similarities between two texts that attempt to correlate Ptolemaic planetary theory to a system of celestial orbs. However, if a text such as ours were written under the direct influence of the *Hypotheses*, one would expect the key elements of that text to appear in our own. They do not.

4. Predecessors and the *hay'ah* tradition

We have just demonstrated that *On the Configuration* does not depend directly upon the *Planetary Hypotheses* ofPtolemy. However, it is quite clear from chapter one of our text [passages 1-5] that Ibn al-Haytham does not regard himself to be the first person to address the problem of the physical description of the heavens. His predecessors in the field have discussed the types of bodies found in the heavens and their arrangement, but their treatment is unsatisfactory in that it does not spell out just how the various motions that have been observed take place on the solid bodies which they themselves have posited.

In this section we shall take note of some of these earlier discussions. In particular, we shall look at two ninth-century treatises that represent, in our view, two distinct genres—one can also call them traditions—of astronomical writing. These are the *Zîj* of al-Battânî and the *Elements of Astronomy* of al-Farghânî. The first is a set of astronomical tables with instructions for their use. It is clearly a specimen of technical astronomical writing, a handbook for professionals. Yet al-Battânî does not ignore cosmology completely, and he even ventures a few remarks of an undoubtedly philosophical character. Al-Farghânî's book, in contrast, is a non-technical exposé. One might have expected that al-Farghânî's audience would have been interested in an expressive and extensive depiction of the celestial regions, but, in any event, such is not to be found in his treatise. Al-Farghânî's presentation falls quite short of Ibn al-Haytham's demand for an explicit enunciation of the way in which the motions of the stars take place on the various spheres. Let us now look at the cosmological discussions found in the books of al-Battânî and al-Farghânî.

At the beginning of chapter 31 of his *Zîj*, al-Battânî gives a fairly detailed account of the structure of the systems of orbs which belong to the planets:

> As for the description of the orbs of the five stars and their situations, which are known by means of their different velocities by the method of proof: Each one of the stars has four orbs similar to the configuration (*hay'ah*) of the orbs of the moon [discussed in an earlier chapter]. The first of them is the parecliptic. Its center is the center of the ecliptic orb, it is symmetric (*mu ᶜtadil*) below it [the ecliptic], and its [parecliptic's] motion is like its [ecliptic's] motion. The second is the inclined orb. Its center is the center of the parecliptic orb and its width is like its width. Its inclination to the parecliptic orb is the greatest amount of latitude of the star to the north or south.

My translation of the last sentence is somewhat idiomatic. Al-Battânî goes on to describe the remaining orbs:

> Inside this orb is another orb which is eccentric vis-à-vis the center of the two [previous] orbs. It is attached (*mutaᶜalliq*) to it, connected to it (*yulâ ṣiquhu*) in one point, which is the point of furthest distance. The equation of anomaly of the center of each star is known from the measure of that which is between the centers of the two orbs [i.e. the eccentricity] by means of the method demonstrated for the moon. Its center moves along (*yajrî ᶜalâ*) this eccentric orb from the point of furthest distance in the direction of the succession of signs with the measure [i.e. velocity] of the mean daily motion of the star in longitude.

Thus al-Battâni provides a generalized description of all the planetary systems which, in fact, offers as much detail as that of Ibn al-Haytham in our own text (setting aside the dirigent orb added by the latter for Mercury). Neither author produces a system as complicated as that of the *Planetary Hypotheses*. The advance evidenced by *On the Configuration* over the account of al-Battânî can be stated simply but correctly as follows. The generalized account offered by al-Battânî is worked out by Ibn al-Haytham separately for each of the planets and is then related to the mathematical constructs, the circles and points, and also to the motions that are described in the *Almagest*.

As we indicated earlier, al-Battânî talks about some physical features of the universe which Ibn al-Haytham chose to exclude from our text, namely planetary distances and sizes. These two topics, though vital to a comprehensive description of the universe, were not deemed necessary for Ibn al-Haytham's more limited task of matching mathematical constructs to physical bodies.

The true nature of the heavenly bodies is, in al-Battânî's opinion, inscrutable [p.182]:

> That which is above these [the four elements] is a fifth nature. Its true character cannot be stated (*lâ yuqâlu fîhâ ḥaqîqah*). The senses do not attain it. Reason cannot comprehend its quality (*lâ yuḥîṭ al-* ᶜ*aql bi-kayfiyyatihâ*).

A somewhat less detailed account is furnished by al-Farghânî. Nevertheless, it is clear that he too is aware of the system of nested celestial spheres. Chapter twelve of his *Elements* is entitled, "On the configuration of the orbs of the stars, their order, arrangement, and distances from the earth." Al-Farghânî begins the chapter with these remarks:

> We shall set forth in this chapter the views of early scientists and that which they have collected concerning it. We say that the number of orbs encompassing all the motions of the stars are eight. Seven of them belong to the seven planets. The eighth, which is the highest, belongs to all the fixed stars; it is the ecliptic orb. The configuration of these orbs is like the configuration of spheres, some of which are inside others. The smallest of them, which is the closest of them to the earth, is the sphere of the moon. The second is Mercury...

Al-Farghânî then lists the seven planets. He continues:

> The orb of the fixed stars is the ecliptic orb. Its center is the center of the earth. The centers of the seven spheres which belong to the seven planets are removed from the center of the earth in different directions.

Note that here al-Farghânî does not elaborate on the position of the eccenters. On the other hand, Ibn al-Haytham puts some effort into the matter (although he does not calculate the eccentricities), especially since most systems have at least two eccenters, namely the deferent center and the equant. In his criticism of his predecessors in chapter one, Ibn al-Haytham observes that one of the things that those people failed to describe adequately was the "deferent spheres with their various centers" [passage 3].

Note also from the title of the chapter that al-Farghânî, like al-Battânî, was concerned with planetary distances. We reproduce here several other chapter titles from the *Elements* which are indicative of al-Farghânî's concern with the physical universe:

> Chapter two: That the heavens are like a sphere and the revolution of all the stars within it is like the revolution of a sphere.
>
> Chapter four: That the sphere of the earth is fixed in the middle of the sphere of the heavens as a center, and its measure compared to the measure of the heavens is as small as the measure of the point to the circle.

Observe that these chapter titles present some of the six cosmological principles enunciated by Ptolemy at the beginning of the *Almagest*.

It is thus quite plausible that the treatises of al-Battânî and al-Farghânî (especially the latter) exhibit the type of discussion which Ibn al-Haytham criticizes in the introduction to our treatise, and Ibn al-Haytham may very well have had those two treatises in mind when he set out to write *On the Configuration*. However, we must always bear in mind that the scientists who flourished within the Islamic orbit were exposed to a wide range of traditions displaying different ideas, varying concerns, and distinctive terminologies. It seems therefore appropriate to round out our discussion by calling attention to some other representatives of *hay'ah* literature.

One of the earliest people to write on this subject is Ya^{c}qûb ibn Ṭâriq, whose *Arrangement of the Orbs* (*Tarkîb al-Aflâk*) was probably composed in 777/8 [Pingree [1968], p. 98; see also Sezgin (1978), pp. 124-27]. Segments of that treatise have been preserved by al-Bîrûnî in his treatise on India and also in the *Astrology* of Ibn Hibintâ ; the material of interest here is found only in the latter. The fragment preserved for us by Ibn Hibintâ is not free of textual problems, some of them quite serious, but it is possible nonetheless to glean from it much information. Ibn Ṭâriq's account is said to derive from the scholars of India. In the system that he describes seven orbs have been created within the orb of the ecliptic. These are the "equal" or "regular" (*mustawin*) orbs. In fact, each "regular" orb is the outermost of a set of nested orbs proper to each planet. These particular orbs account for the daily, westerly motion. We find also a clear statement of the mechanics involved: "They [the "regular" orbs] neither turn nor move, but rather the ecliptic orb turns them along with itself from east to west" [p. 26]. Each planet is said to have six orbs, that is, five orbs in addition to the "regular" orb. Their complete enumeration, however, is unclear; an orb of the node, orb of the center (!) (*falak markaz*), and small orb (no doubt the epicycle) are clearly mentioned. The moon has three orbs: regular, node, and apogee; and the sun has two, regular and apogee.

Ibn Ṭâriq's terminology is also very deserving of our attention. We have already noted two unusual terms, *falak mustawin* and *falak markaz*. Neither of these terms figure in the treatise of Ibn al-Haytham. Another item of more purely philological interest is ibn Ṭâriq's term for "against the order of the signs": *mankûsan ^{c}alâ ta'lîf nuks (!) al-burûj min al-ḥamal 'ilâ al-ḥût* [p. 30].

P. Duhem [1914, p. 126] calls our attention to two passages from *The Guide of the Perplexed* of Moses Maimonides (d. 1204) in which reference is made to a treatise of Thâbit bin Qurra, a ninth-century astronomer, which treats of the physical structure of the universe. The first passage comes from book II, chapter 24 [vol. 2, pp. 324-325] of the *Guide*:

> Hence necessity obliges the belief that between every two spheres there are bodies other than those of the spheres. Now if this be so, how many obscure points remain? Where will you suppose the centers of those bodies existing between every two spheres to be? And those bodies should likewise have their own particular

> motion. Thabit has explained this in a treatise of his and has demonstrated what we have said, namely, that there must be the body of a sphere between every two spheres. All this I did not explain to you when you read under my guidance, for fear of confusing you with regard to that which it was my purpose to make you understand.

The other passage comes from book III, chapter 14 [vol. 2, p. 457]:

> For the thickness of the bodies of the spheres has only been demonstrated with respect to its minimum possible dimensions, as is made clear in the Epistles concerning distances. Similarly it is possible to determine exactly the thickness of the bodies that, according to Thabit, reasoning forces us to admit between every two spheres. For in these bodies there are no stars from which an inference can be drawn.

To the best of my knowledge, these works of Thâbit have yet to be identified. A fragment published just now by R. Morelon [1988] does contain a discussion on the part of Thâbit of a problem of celestial physics, which, however, is not related to the problem of the arrangement of the spheres.

From the point of view of style there is no question that Ibn al-Haytham's *On the Configuration* bears some resemblance to al-Farghânî's *Elements*. But can we state conclusively that the treatises mentioned and criticized in the opening passage of *On the Configuration* refer, in fact, to the book of al-Farghânî and whatever other treatises of a similar nature may have been known to Ibn al-Haytham? Was Ibn al-Haytham mindfully contributing to a distinct tradition in astronomical literature, which we call the *hay'ah* tradition? We would at present somewhat hesitatingly proffer the following answers. Ibn al-Haytham's criticisms of his predecessors takes in all those who have written on the subject, including, for example, al-Battânî, for no one has yet provided a satisfactory three-dimensional account of the heavenly motions. On the other hand, Ibn al-Haytham chose to set forth his own presentation in a work similar to that of al-Farghânî, perhaps because he surmised that the audience for such an account comprised people with an interest in natural science, philosophy, or learning in general, rather than professional astronomers.

Ibn al-Haytham himself has left us an important clue concerning this matter, which has been discovered by M. Schramm [1963, pp. 66-67]. In his own listing of his literary output, Ibn al-Haytham placed *On the Configuration* among his works in natural science and metaphysics (*al-ᶜulûm al-ṭabîᶜiyyah wa-l-ilâhiyyah*) rather than, as one might expect, the mathematical sciences (*al-ᶜulûm al-riyâḍiyyah*). According to Schramm, Ibn al-Haytham has made a purely formal classification: since *On the Configuration* contains no mathematical proofs, it cannot be classified as mathematical. Schramm's point is well-taken; yet it seems to me that there may be something more to this. It may just be that Ibn al-Haytham is hinting at the target audience of this treatise. That is to say, *On the Configuration* is aimed at those whose chief interests lie in the realms of natural science and metaphysics, and the lack of technical discussion is due as much to the presumed limited capacity of this audience as it is to the formal structures of those two disciplines.

In any event, the history of the *hay'ah* tradition has yet to be written; nor are we quite ready to write it, although scholarship has taken some significant strides in the last few years. Much groundwork needs to be done in the intellectual history of medieval Islamic civilization. J. L. Kraemer [1986] has contributed two excellent studies which combine a wide perspective with a concern for detail; works of this sort surely facilitate more specialized investigations into the histories of the various branches of learning. For the history of astronomy we especially need more studies which focus on the interest shown and importance attached to astronomy within the various groups of intellectuals—not to speak of an analysis of the astronomical literature itself. F. J. Ragep [1982] has recently given research on this problem a very good thrust; he has as well some cogent remarks on the role of Ibn al-Haytham's *On the Configuration* in congealing the *hay'ah* tradition.

For now we may summarize as follows. *On the Configuration* of Ibn al-Haytham and the *Elements* of al-Farghânî are two specimens of what can be more or less designated a distinct tradition in Islamic astronomy, the *hay'ah* tradition. The main characteristics of this tradition are a concern for the three-dimensional structure of the heavens and the physical principles of planetary motion, all within the context of a presentation of the then-accepted body of astronomical knowledge and its apparatus, usually—though, perhaps, not always—in the form of a non-technical exposé. We cannot here discuss the developments in this tradition posterior to (and quite probably influenced by) Ibn al-Haytham. Two aspects of these later developments may, however, be suggested. These are: the mathematical ingenuity of some later writers, especially Naṣîr al-Dîn al-Ṭûsî, his colleagues and followers; and the concerns of such figures as Maimonides and Ibn Rushd (Averroes) who were primarily philosophers but were fully versed in the technical details of Ptolemaic astronomy. It appears that the Andalusian philosophers and the Eastern savants were pursuing quite separate and independent paths; only the latter were responding to the criticisms that Ibn al-Haytham raised in the *Doubts* [Langermann, 1990]. However, in a recently published fragment Ibn Bâjja, the eleventh-century Spanish philosopher who greatly influenced both Maimonides and Ibn Rushd, sharply criticizes the *Doubts* [Ma^{c}ṣûmî, 1983, p. 78; the passage of interest had already been cited by Pines (1964), n. 10]; this indicates that the Andalusians were quite aware of Ibn al-Haytham's work almost as soon as it appeared. The intellectual setting in Spain, recently discussed by Sabra [1984], thus seems somewhat more complex than previously thought.

As we have just stated, a full description of the reception and assimilation of *On the Configuration* is beyond the scope of this study. However, I would like to present a passage from one of the writings of Levi ben Gerson (1288-1344), one of the outstanding Jewish philosophers and astronomers of the medieval period. This passage is instructive concerning some of the unexpected ways that our text figures in medieval discussions of cosmology, and, from a wider perspective, the episode which we are about to recount is illuminative of some of the ironies of intellectual history.

The passage of interest is found in Levi's supercommentary to Ibn Rushd's short commentary to *De Coelo* [Steinschneider (1893), p. 127], and the problem at hand is as follows. Cosmological doctrine as understood and accepted by Ibn Rushd and by Levi teaches that the stars move in their orbits with an unceasing and unchanging motion. It is further known that no forced or violent motion can be unceasing. Yet the planets exhibit motions in two opposite directions, namely, an easterly proper motion and the westerly daily motion; and since one body cannot possess two intrinsic contrary motions, it appears that at least one of these motions must be forced. Ibn Rushd summarizes the dilemma thus [*Rasâ'il* , p. 59; Hebrew translation, Ms. Paris, Bibliothèque Nationale Heb. 935, f. 60b]:

> Therefore, it has been assumed that each of the planets has two orbs. One of them is the orb by means of which

> it moves from east to west, following the motion of the encompassing [orb]. The second is its orb which turns it from west to east, and it is joined to the other orbs and follows them.

Ibn Rushd, it would seem, is referring (somewhat sketchily) to the arrangement of orbs that is discussed by Aristotle (*De Coelo* II, 12, 293a5 ff.; and, in more detail, *Metaphysics* XII, 8, 1073b17 ff.). Levi, however, understands differently, specifically, he gathers that the anonymous "it has been assumed" refers to none other than Ibn al-Haytham. Although Levi's work is generally called a supercommentary, it is actually a paraphrase *cum* commentary; only the more lengthy excurses are prefaced by the words "Levi said." In the passage below (translated from MS. Vatican, Heb. 342, f. 154b), the comments which Levi has inserted into his paraphrase have been italicized:

> Therefore, it has been assumed that each of the planets has two orbs. One of them is the orb by means of which it moves from east to west, following the motion of the encompassing [orb], *just as Ibn al-Haytham has arranged this in the configuration (t ekhunah) which he has set down. However, others have not assumed any orb for this [westerly motion] other than the encompassing [orb]. All this requires an investigation (ciyyun), but it is not relevant to this science* . The second is its orb which turns it from west to east, and it is joined to the other orbs and follows them.

It is quite clear to me that Levi has in mind *On the Configuration*. Indeed, the Hebrew term *t ekhunah*, which may mean configuration or model, forms part of the title of the Hebrew translation of our treatise. Nowhere in *On the Configuration* does Ibn al-Haytham appear perplexed by the existence of contrary motions in the heavens, that is, easterly and westerly motions, nor does he ever raise the question of the force which powers the planetary motions. (Also of great interest is Levi's remark that the investigation of the models which are supposed to take care of these problems does not properly belong to *De Coelo* and its commentaries.) Yet Levi seems quite certain that the models that were proposed by Ibn al-Haytham were meant to answer this very dilemma; he interprets the solutions of Ibn al-Haytham wholly within the context of the problems which were of concern to *him*. As a matter of fact Ibn al-Haytham has not assigned to each planet an orb which moves with the daily motion, although, in general, he does suppose orbs other than the encompassing orb which move to the west [passage 47]. This simple error leads me to suspect that Levi had only second-hand knowledge of Ibn al-Haytham's treatise.

Both Ibn Rushd and Levi ben Gerson reject the two-orb solution, since one of the motions, either the easterly or the westerly, must still be considered to be forced. It is somewhat ironic that, in the end (at least in the context of their analysis of *De Coelo*), both opt for the analogy to an animal which, on the strength of one unified force, moves its various limbs in different directions. For, in fact, the simile of the celestial animal (*al-ḥayawân al-falakî*) is invoked by Ptolemy in Book II of the Planetary Hypotheses! [Goldstein (1967), p. 41]. It should also be borne in mind that both Ibn Rushd [Sabra (1984), pp. 139-140] and Levi [MS. Vatican, Heb. 342, f. 94b] were familiar with the *Hypotheses*. We have already seen how some modern scholars have mistakenly asserted that *On the Configuration* is but a reworking of the *Hypotheses*. Here we encounter a medieval scholar, Levi ben Gerson, who rejects a different misinterpretation of *On the Configuration* and chooses instead (following the lead of Ibn Rushd) an idea which is truly found in the *Hypotheses*!*

5. The Medieval Translations

Hebrew Translations

We know of three Hebrew versions. Our text was first translated by Jacob ben Machir ibn Tibbon sometime between 1271–1275. (The years 5031, 5033, and 5035 A.M. are found in different manuscripts). A second translation was accomplished by Shlomo ibn Paṭer, a medical doctor, at Burgos in 1322, at the request of one Jacob ben Meir. A third version—perhaps a retranslation, perhaps a recension of ibn Tibbon's translation that was checked against the

---

* It must be reemphasized that this solution is given in the context of the two scholars' discussion of *De Coelo* ; their own personal cosmologies are quite another matter. Levi's own investigations combined thorough and original astronomical research with philosophical reflection; see the most recent study of B. R. Goldstein [1985] in its entirety and the literature cited therein. There is no comparable study of Ibn Rushd. The paper of F. J. Carmody [1952] is sketchy and not without errors; Sabra [1984] contributes some cogent remarks and substantive hints which may provide direction for any future project.

Arabic original—is extant in one manuscript. Yet a fourth version, comprising, as it appears, only a few chapters, has been included as part of a larger Hebrew work. These last two translations will be discussed in detail below.

In the apparatus to our English translation we have cited the variants found in the manuscripts listed by Steinschneider [1893, p. 560], according to the following key:

H1. Vatican, Hebrew 399, ff. 1-87. Figures: zodiacal circles; sun; moon; Mercury; Venus (incomplete). Translator's introduction.

H2. Rome, Casanatense 3082, ff. 1-21. No figures.

H3. Parma, Bibliotheca Regia, De Rossi 568 (2466), ff. 2-33. Figures: division of ecliptic; another detailed figure of the ecliptic; Venus; superior planets; what looks like a generalized view of planetary latitudes as seen on the highest orb.

H4. Paris, Bibliothèque Nationale Heb. 1022, ff. 49-83. No figures.

H5. Paris, Bibliothèque Nationale Heb. 1021, ff. 31-48. No figures.

H6. Paris, Bibliothèque Nationale Heb. 1031, ff. 83-100. Figures: zodiacal circles; sun.

H7. Paris, Bibliothèque Nationale Heb. 1035, ff. 1-37. Figures: sun; moon; Mercury; Venus; Venus (second, incomplete figure); superior stars (incomplete).

H8. Vienna, Oesterreichen Nationalbibliothek, Cod. Heb. 66, ff. 117-154. No figures. Translator's introduction.

H2,3,4,5, and 7 are copies of Jacob ibn Tibbon's translation. H6, 8 are copies of the translation of Ibn Paṭer. In the years since the preparation of our edition, quite a number of additional copies have been catalogued by the staff of the Institute of Microfilmed Hebrew Manuscripts, Jewish National and University Library, Jerusalem. They are:

9. Madrid, Real Academia Historia 8, ff. 194a(199a)-216b(221b). [Ibn Paṭer]

10. Mantua, Jewish Community no. 4, ff. 134b-157b [Ibn Tibbon]

11. Parma, Regia 3824, ff. 96b-136a [Ibn Tibbon]

12. Dresden, Eb 384, ff. 23a-39b [Ibn Tibbon]

13. Cambridge, Add. 650, 1, ff. 2a-16a [Ibn Tibbon]

14. Oxford, Bodley Michael 400 [Neubauer 2006], ff. 148a-149a. Chapter 1 and part of chapter 2. [Ibn Tibbon]

15. Oxford, Bodley Michael 144 [Neubauer 2027], ff. 94a-b, beginning only [Ibn Paṭer]

Unless indicated otherwise, these are all complete copies. Two other manuscripts are listed by Steinschneider [1893, p. 560]. They are: (**16**) Leningrad, Firkovitch 1879, n. 13; and one in a private collection, which we assume to be that which is now at Cambridge (no. 13). A few days before submitting the final version of this book we received, though the gracious cooperation of the Saltykov-Shchedrin Library, a film of I Firkovitch 354, an apparently complete copy of the Ibn Tibbon translation, executed at Constatinople by the Karaite Isaac ben Moses. I do not know if this is the same as the copy noted by Steinschneider. There exists a copy also at (**17**) Moscow, Ginzburg 1122; this is also unavailable. Just several weeks before the completion of this book I was informed that two additional copies are located in the collection of the Oriental Institute at Leningrad: (**18**) C76, ff. 87-105 and (**19**) B70, ff. 8-30. Thus we know of a total of nineteen (or twenty) copies, most of them complete and provided with figures. A twentieth copy, to be discussed below, is incorporated as part of an independent Hebrew treatise. We have also discovered two Genizah fragments of the Arabic original, written in Hebrew characters. These are discussed in the following section along with the other Arabic manuscripts.

M. Steinschneider [1881, p. 725] first identified the translations of Ibn Tibbon and Ibn Paṭer. Regarding their relative merits he concluded:

> . . .celle de Salomo [ibn Pater] est moins arabistique et un peu plus intelligible.

We concur in this conclusion. For example, Ibn Paṭer uses the familiar *naga* ^{c} for the Arabic *mâssa* (to touch), whereas Ibn Tibbon employs the Hebrew cognate *m^{e}mashshesh*. Similarly, Ibn Paṭer translates *ikhtalafa* (to be different) by *hishtanneh*, whereas Ibn Tibbon again prefers the cognate form *hitḥallef*.

It also seems to be the case that Ibn Paṭer was working with an Arabic text which, in a few places, possessed a somewhat better reading than the Arabic manuscript that we have seen. (See note 6 to passage 213 and note 4 to passage 266; note that both of these passages are found in chapter ten, which is missing from one of our two Arabic manuscripts). Ibn Paṭer was also a bit more precise in his choice of technical terminology. For example, he employs *^{c}iqqum* as a special term for the "slant" of the epicyclic diameter. The corresponding Arabic word, *mayl*, is also used for "declination", and, unlike Ibn Paṭer, Ibn Tibbon chose to use the same Hebrew term, *n^{e}ṭiyyah*, for both "slant" and "declination". (See also passage 336, note 9.)

The Vatican manuscript, which we have designated H1, stands apart from all of the others. Steinschneider [1881, p. 723], who, it should be noted, seems to have been working with extracts only, attributes this translation to Ibn Tibbon. While the language of this manuscript is indeed very close to that of Ibn Tibbon's translation, a close analysis shows that this copy represents an independent effort to understand the Arabic original. We shall now discuss some of the features of this manuscript in detail.

Let us first turn to the translator's introduction, which is unique to this manuscript. After much poetry in praise of knowledge, the translator tells us that he has studied no science (*ḥokhmah*) nor has he studied with any scholar. On his own he collected many books and devoted himself to study. One day he was visited by a friend from a distant land who was anxious to learn astronomy. As the two discussed astronomical matters, the translator sensed that his own appetite for knowledge in that field was increasing. The visitor expressed dissatisfaction with the work of al-Farghânî and convinced the translator to tackle our text. The writer of H1 says that although he is not a professional translator (*me-anshei ha-ha^{c}ataqah*), he undertook this effort, partly in order to improve his command of Arabic.

Now we are not very well-informed concerning Jacob ibn Tibbon's biography [see Vernet,1975], nor is the translator's introduction to H1 transparent in all places. Nevertheless, we know enough to call into doubt the identification of Ibn Tibbon with the writer of H1. Jacob ibn Tibbon was born about 1236 and his translation of our text dates, as we have seen, from the early 1270's. We know that he studied medicine at Montpelier, and he may have spent a year in Gerona studying with the great Rabbi Moses ben Naḥman (Naḥmanides). Moreover, he was the fourth generation of a family of famous translators and scholars. Thus the picture of a man left entirely to his own resources for his education, and who expressly denies being a professional translator, does not fit well with what one might expect of Jacob ibn Tibbon. Steinschneider [1881, p. 723], perhaps troubled by all of this, took this introduction to be motivated by "une modestée, peut-être affectée."

In any event, there is much internal evidence to distinguish H1 from the other translations, as the reader may note in our notes to the English translation. By way of example, let us call particular attention to the variants noted for passages 72 (n. 2) and 175 (n.5), where all the Hebrew translations save H1 are missing part of the Arabic original. Also interesting are a few places in the text where H1 uses several words to translate the same Arabic phrase, which we have noted in the apparatus to passages 118 (n. 3) and 315 (n.3), as if to suggest that neither word captures the full and exact sense of the Arabic. This indeed seems to be a not unusual feature of medieval translations in general. In one instance (passage 359, n. 3) H1 displays, one after the other, both the Ibn Tibbon and Ibn Paṭer translations of a particular sentence. In passage 309 (n.8) H1 inserts an Arabic term (*mayl*) in the middle of his Hebrew sentence. Noteworthy also is the variant to passage 173 (n.5), where H1 reproduces a quirk found in one of the Arabic manuscripts. In most places, however, H1 follows the translation of Ibn Tibbon. Our feeling is that the writer of H1, rather than executing a completely new translation, wished instead to improve upon Ibn Tibbon's translation by closely comparing it with the Arabic original. As a matter of fact, most of the manuscripts bear a number of unique readings, something which may indicate that the various scribes and/or students of our text felt free to make minor emendations, which, in their opinion, would make the meaning more clear.

Another partial translation of *On the Configuration* exists in a manuscript at the Bibliothèque Nationale at Paris, MS. Heb. 1045. This text affords us a look at another aspect of medieval scholarship, namely the unacknowledged incorporation on the part of one author into his own treatise of segments of the work of another; we should add that medieval perceptions of plagiarism were quite different from those of the present time, and we should be accordingly restrained in our pronouncements. [See von Grunebaum, 1944.] The last item found in BN Heb. 1045 (ff. 215a-221b), hitherto unnoticed [cf. Zotenburg,1866, p. 190], is a fragment of a Hebrew work. Unfortunately, neither the beginning nor the end of the treatise has been preserved, and there is no indication in the surviving portions as to the author's name, date, or location. With some caution we may venture the guess that the work was written in Spain or southern France in the thirteenth or fourteenth centuries. It can, however, be said with certainty that the book is the work of a Jewish author, who argues in favor of the study of astronomy. On f. 217a he complains, "I have seen many of this generation who are immersed in Torah, Mishnah, and religious instruction, whose custom is not to inquire into these sciences." At least two chapters from *On the Configuration*, those dealing with the orbs of the sun and the moon (chapters nine and ten), have been included as part of this work (ff. 219a-221b). The chapter on the sun is preserved in its entirety; that on the moon breaks off towards the end of passage 252, at the point where the fragment itself terminates. I am quite satisfied that these portions of *On the Configuration* have been translated by the author directly from the Arabic. His Hebrew text differs from both the Ibn Tibbon and the Ibn Paṭer translations. In particular, this writer seems to prefer the technical vocabulary of Abraham ibn Ezra, whose distinctive terminology has been noted by G. Sarfatti [1969, p. 145]. He uses *muṣaq* for "center" and *seden* for "pole," though he also employs *merkaz* and *qoṭev*, the more common terms, which are also the terms utilized in the other translations. This new source offers further evidence for the popularity and importance of Ibn al-Haytham's treatise among Jewish intellectuals.

To judge both from the number of translations and from the number of surviving copies, *On the Configuration* was extremely popular among the European Jewish communities situated along the western Mediterranean basin. Scholars have traditionally emphasized the role of Hebrew translators and translations as intermediaries between the Arabic-speaking and Latin-writing environments during the Middle Ages. However, it is clear to us that in most cases, scientific and philosophical works were translated into Hebrew primarily in response to internal, specific needs.

Ibn al-Haytham's *On the Configuration* was translated, in whole or in part, a total of four times. One of the translations was executed by a scion of the most famous of all families of translators, the Ibn Tibbons, who played a major role in stimulating interest in science and philosophy, particularly among the Jews of Provence. It is worth noting in this context that members of this family, including Jacob, the translator of our treatise, produced original Hebrew works as well. To judge by its vocabulary, we may say that the second translation of our text, that of Ibn Paṭer, was carried out at least in part in response to stylistic and nationalistic demands for versions of texts that were more truly Hebrew, that is, free of Arabisms. Yet a third translation, or revision of an earlier translation, was made in response to the demands of an anonymous patron, and a fourth translation was undertaken in order to supply a Hebrew treatise with some chapters on astronomy. The profusion of manuscripts indicates a real popularity for our text.

Can we venture a guess as to the reasons for this popularity? It seems to me to be connected to the burgeoning of Jewish interest in science and philosophy beginning in the thirteenth century which, in turn, may be attributed in good measure to the reception of the Maimonidean philosophy and the absorption of its principal ideas. (In fact, most of the Hebrew manuscripts seem to date from the fifteenth century; I have as yet no explanation for this.) Maimonides makes it clear—particularly in the opening remarks of his *Guide*—that the path to religious philosophy is open only to those who have first mastered the sciences, especially astronomy. Now it was probably beyond the ability of most seekers of wisdom to fathom the technicalities of the *Almagest*. Ibn al-Haytham's *On the Configuration*, and such related texts as the *Elements* of al-Farghânî (which also survives in a large number of Hebrew manuscripts; see Steinschneider [1893, pp.554-556]), fit the bill much better. Indeed, we have already suggested that one of the main goals of the treatise may have been to provide just such a non-technical exposé.

Latin Translations

We have listed in the apparatus to our English translation variants from one Latin translation of our text. This basic reference text is:

L1. Madrid, Biblioteca Nacional 10059, ff. 37a-50ra. No figures.

L1 is the only complete version of *On the Configuration* which has been published to date, having been transcribed by J. Millás Vallicrosa and included as an appendix to his *Los traducciones orientales en los manuscritos de la Biblioteca Catedral de Toledo* [1942, pp. 285-312]. Millás judges the manuscript of this anonymous translation to date from the late thirteenth century or early fourteenth century. On the basis of a comparison of L1 with a Hebrew translation (which one is not specified), Millás suggests that L1 is a literal translation [1942, pp.206-207]. In this he is quite correct. L1 is, however, missing chapter one (the introduction).

A second Latin version of our text is found in a manuscript at Oxford, Canon Misc. 45, ff. 1a-57a, 42 figures. This is is much more a reworking of our text than a translation, with regard to both contents and organization. It forms part of the corpus of Arabic scientific writings that were made available in Latin under the patronage of Alfonso X (the Wise) of Castille (1221-1284). We have not listed the very numerous and very lengthy variants from this manuscript. However, the distinctive terminology of this translation has been included in the lexicon; it is labeled there **L2**.

There exists yet a third Latin translation, Vatican 4566, ff. 1-43. This translation, Steinschneider [1881, p. 724] tells us, was made by Abraham of Balmes (d. 1523) for Cardinal Grimani from the Hebrew translation of Jacob ben Machir ibn Tibbon. Steinschneider notes further [p. 729] that this manuscript is of some philological value but, on the whole, this copy leaves much to be desired. It appears that Balmes had some trouble with the Hebrew text, especially chapter one, and that the astronomical subject matter as well may have been foreign to him. In fact, Steinschneider observes, Balmes' other translations, which consisted mainly of Islamic philosophical works, are also unsatisfactory.

This manuscript contains a "Prohemium," apparently a translation of chapter one, which is quite difficult to comprehend. (The "Prohemium," along with some other selections from this translation, have been published with notes by Steinschneider [1881, pp. 732-736].) In fact, Duhem [1914, p. 122], who based his appraisal of our text upon several passages that were copied for him from this manuscript, calls this section "un extraordinaire galamitias". Nevertheless, the sparse quotations from this manuscript which Duhem has translated into French do conform to the Arabic text.

Certain abbreviations which are employed in this copy also pose a major problem. Steinschneider [1881, p. 729, n.2] remarks:

> Il me faut remarquer que la solution des abbréviations n'était pas toujours facile et qu'il y aura peut-être encore quelque méprise.

We have looked at this manuscript. However, in view of the difficulties noted above and, in particular, given the fact that we have made a thorough study of the Hebrew texts upon which this translation is based, we have not included this manuscript in our apparatus.

6. Our Edition and Translation

The Arabic text was established on the basis of the two Arabic manuscripts described below, copies of which were available to us at the time of the preparation of our edition. An additional manuscript, located in Morocco, has since been listed by F. Sezgin [1978, p. 255]. We were unable to obtain a copy of that. Sezgin also records [p. 294] an additional small fragment, Princeton 1168, ff. 62b-64a; we have inspected a photocopy of those pages, and they do not belong to our treatise. We have also discovered two fragments in the Cairo Genizah, Arabic in Hebrew characters. These are described at the end of this section.

We have divided the text into a number of consecutively numbered "passages". The English translation follows the same system of division. The existence of variants in the Hebrew and/or Latin translations is indicated by numerals in superscript to the English translation; the variants are listed in the apparatus which follows each chapter. On a number of occasions we have felt the need to add brief explanatory footnotes. These are indicated by an asterisk (*). We have also compiled an Arabic-English-Hebrew-Latin lexicon for some key words.

The two Arabic manuscripts are:

Arabic Y. London, India Office, Loth 734, ff. 101-116. Average 36 lines per page. No figures or chapter titles.

Arabic K. Kastamonu 2298, ff. 1-43. Average 17 lines per page. One figure and most chapter titles are given.

Arabic Y was first brought to the attention of scholars by Steinschneider [1883], who found it described in Otto Loth's *Catalogue of the Arabic Manuscripts of the Library of the India Office* (London,1877), page 212. This same manuscript contains some seventeen other works of Ibn al-Haytham. Steinschneider [1883, p. 506] describes this manuscript as follows:

> Le manuscrit est très bien écrit, muni des signes pour les voyelles en cas de ambiguité ou de doute, et en général bien correct, mais les caractères sont si petits, que je fus obligé, en égard de mes yieux fatigués, de restreindre mon examen plus que je n'aurais pas souhaité.

Thus Steinschneider did not examine the whole text, limiting himself to a few passages. Nevertheless, we generally agree with Steinschneider's opinion. The manuscript is well-written. However, we did not find the vocalization to be that consistent. Kohl [1923, p. 144] adds that this manuscript dates from the sixteenth century. Arabic Y is also the only version in any language which contains the appendix. Although no chapter titles are provided, there are indentations corresponding to the division found in the other manuscript. There are no spaces where figures might have been drawn in.

The existence of the second manuscript, Arabic K, was announced by A. I. Sabra [1972, p. 205]. As Sabra notes, this manuscript is incomplete. The most serious omission is that of two entire chapters, nine ("The Orb of the Sun") and ten ("The Orb of the Moon"). The last paragraph of chapter eight, which introduces the second part of our treatise, is also missing. There is another, shorter omission in chapter three, stretching from the middle of passage 55 to the middle of passage 62. Pages 14a, 14b, and 30a are written in a different hand than the rest of the manuscript. Some of the folios have been misbound. Chapter thirteen does not begin right after chapter twelve (f. 35a), but rather in the middle of chapter fifteen (ff. 41a-42a). The rest of chapter thirteen is found on ff. 35a-b. The only figure in this manuscript is found on page 24b, at the end of chapter eight. It depicts the equinoctial, ecliptic, horizon, and other circles.

The colophon to Arabic Y contains the usual praises but tells us nothing about the text. On the other hand, the colophon to Arabic K reads:

> This book was written from the text transcribed from the text of Abû al-Qâsim al-Samîsâṭî, in his hand. He said that he copied it from a text in the hand of the author the book, the *shaykh* Abû ᶜAlî al-Ḥasan bin al-Ḥasan bin al-Haytham. He collated it from the beginning to the end in Rajab of 476 [Sept.-Oct. 1083].

This collation thus took place some 43 years after the author's death. A second colophon, which may be in a different hand, reads:

> The text from which this text was written was compared with the previously mentioned original text, which is in

> the hand of the *shaykh* Abû ᶜAlî ibn al-Haytham, and corrected. Praise to God, Lord of the World. It was written in Rajab of the previously mentioned year.

These corrections are given in the margin, marked occasionally by the letters *ṣḥḥ*. In one instance Arabic K cites a different reading which is explicitly attributed to another version (see passage 52, note 2 to the English translation).

The task of editing the Arabic text was made much easier by the excellent state of both manuscripts. The overwhelming majority of variants noted in the edition stem from points of grammar, especially gender agreement. The few substantial variants have been translated and exhibited in the apparatus to the English translation.

Finally we note the Genizah fragments mentioned earlier. They are:

(1) **T. S. Arabic 29.19** and **29.95**

These fragments clearly derive from the same manuscript. Each consists of one leaf (two sides). They are written 25 lines to the page, about 8 words to the line, in an early Eastern hand. No. 19 contains chapter eleven, from the middle of passage 297 to the middle of passage 306. No. 95 contains chapter twelve, passage 331, through chapter thirteen, middle of passage 345.

(2) **T. S. Arabic 29.9**

One leaf (two sides), a trifle damaged but still quite readable. Twenty lines are found on the page, about 5 or 6 words to the line. This is a fragment of chapter ten, running from the end of passage 248 to the middle of passage 250.

References

Primary Sources

al-Battânî, *Zîj*

Al-Battani sive Albateneii Opus Astronomicum, ad fidem codicis escurliensis arabice editum, latine versum, adnotionibus instructum, vol. 1, ed. C. A. Nallino (Milan, 1899).

al-Farghânî, *Elements*

Muhammedis Fil. Keteris qui vulgo Alfraganus dicitur, Elementa Astronomia, Arabice et Latine, ed. Jacob Golius (Amsterdam, 1699).

Ibn Bâjja, *Rasâ'il*

Rasâ'il Falsafiyyah li-abî Bakr bin Bâjah, ed. Jamâl al-Dîn Ma^{c}ṣûmî (Beirut, 1983).

Ibn al-Haytham, *Doubts*

Al-Shukûk ^{c}alâ Baṭlamyûs, ed. A. I. Sabra and N. Shehaby (Cairo, 1971).

Ibn Hibintâ, *Astrology*

Al-Mughnî fî aḥkâm al-nujûm. The Complete Book of Astrology. Facsimile edition, F. Sezgin, 2 vols., (Frankfurt a. M., 1987).

Ibn Rushd, *Rasâ'il*

Rasâ'il Ibn Rushd, (Hyderabad, 1948).

Maimonides, *Guide*

The Guide for the Perplexed, trans. S. Pines, 2 vols., (Chicago and London, 1963).

Ptolemy, *Almagest*

Ptolemy's Almagest, new annotated translation by G. J. Toomer (New York-Berlin-Heidelberg-Tokyo, 1984).

Ptolemy, *Planetary Hypotheses*

J. Heiberg, *Claudii Ptolemai opera qui estant omni*, vol. 2, *Opera astronomica minora* (Leipzig, 1907); translation from Arabic of book II of the *Hypotheses* by L. Nix .
Bernard R. Goldstein, "The Arabic Version of Ptolemy's *Planetary Hypotheses*," *Transactions of the American Philosophical Society*, new series 57, no. 4 (1967).

Secondary Sources

Carmody, F. J.

[1952] "The Planetary Theory of Ibn Rushd," *Osiris* , vol. 10, 556-86.

Corbin, Henri

[1986] *Histoire de la philosophie islamique* (new edition, Gallimard, Paris).

Duhem, Pierre

[1914] *Le Système du monde,* vol. 2 (Paris).

Goldstein, Bernard R.

[1985] *The Astronomy of Levi ben Gerson (1288-1344)* (New York-Berlin-Heidelberg-Tokyo).

Grant, Edward

[1978] "Cosmology," in D. Lindberg (ed.), *Science in the Middle Ages*, 265-302.

von Grunebaum, G. E.

[1944] "The Concept of Plagiarism in Arabic Literary Theory," *Journal of Near Eastern Studies*, 234-53.

Hartner, Willy

[1965] "Falak," *Encyclopaedia of Islam*, second edition, vol. 2, 761-63.

Jardine, Nicholas

[1982] "The Significance of the Copernican Orbs," *Journal for the History of Astronomy*, vol. 13, 168-94.

Kohl, Karl

[1923] "Über den Aufbau der Welt nach Ibn al-Haitam," *Sitzungsberichte der physikalisch-medizinischer Sozietät in Erlangen*, 54-55 (1922-23), 140-179.

Kraemer, Joel K.

[1986] *Humanism in the Renaissance of Islam* (Leyden).

[1986] *Philosophy in the Renaissance of Islam* (Leyden).

Langermann, Y. Tzvi

[1982] "A Note on the Use of the Term *orbis* (*falak*) in Ibn al-Haytham's *Maqâlah fî Hay'at al- ᶜâlam*," *Archives internationales d'histoire des sciences*, vol. 32, 112-113.

[1990] "The 'True Perplexity': The Guide for the Perplexed II,24," in J. L. Kraemer (ed.), *Perspectives on Maimonides* (Oxford), 193-208.

Millás Vallicrosa, José M.

[1942] *Los traducciones orientales en los manuscritos de la Biblioteca Catedral de Toledo*, (Madrid).

Morelon, Régis

[1988] "Les deux versions du traité de Thâbit b. Qurra 'Sur le mouvement des deux luminaires," *Mélanges. Institut Dominicain d'Études Orientales de Caire*, vol.18, 9-44.

Nallino, C. A.

[1921] "Sun, Moon, and Stars (Muhammadan)," Hastings' *Encyclopaedia of Religion and Ethics*, vol. 12, 88-101.

Neugebauer, Otto

[1975] *A History of Ancient Mathematical Astronomy*, 3 vols., (Berlin-Heidelberg-New York).

Pines, Shlomo

[1946] "The Function of Astronomical Theories according to Ibn al-Haytham," [Hebrew] *Tarbiz*, vol. 17, 101-104.

[1964] "Ibn al-Haytham's Critique of Ptolemy," *Proceedings of the Tenth International Congress of the History of Science at Ithaca 1962*, (Paris) 547-550.

Pingree, David

[1968] "The Fragments of the Works of Ya^{c}qûb ibn Ṭâriq," *Journal of Near Eastern Studies*, vol. 27, 97-125.

Ragep, F. Jamil

[1982] *Cosmography in the "Tadhkirah" of Naṣîr al-Dîn al-Ṭûsî*,.[Doctoral Dissertation, Harvard University, History of Science].

Sabra, A. I.

[1972] "Ibn al-Haytham," *Dictionary of Scientific Biography*, vol 6, 189-210.

[1979] "Ibn al-Haytham's Treatise: Solution of Difficulties Concerning the Movement of Iltifâf," [Arabic text, English synopsis], *Journal for the History of Arabic Science*, vol. 3, 392-422.

[1984] "The Andalusian revolt against Ptolemaic Astronomy," in E. Mendelsohn (ed.), *Transformation and Tradition in the Sciences*, (Cambridge, U.K.), 133-53.

Sarfatti, G. B.

[1969] *Mathematical Terminology in Medieval Hebrew Scientific Literature*, [Hebrew], (Jerusalem).

Schramm, Matthias

[1963] *Ibn Al-Haytham's Weg zur Physik*, (Wiesbaden).

Sezgin, Fuat

[1978] *Geschichte des arabischen Schrifttums* , vol. 6 , (Frankfurt a. M.).

Steinschneider, Moritz

[1881] "Notice sur un ouvrage astronomique inédit d'Ibn Haitham," Boncompagni's *Bulletino di bibliografia e di storia della scienze matematiche e fisiche*, vol. 14, 721-736, and Hebrew appendix.

[1883] "Supplément à la Notice sur un ouvrage astronomique inédit d'Ibn Haitham," Boncompagni's *Bulletino di bibliografia e di storia della scienze matematiche e fisiche*, vol. 16, 505-513.

[1893] *Die hebraeischen Uebersetzungen des Mittelalters und die Juden als Dolmetscher*, (Berlin).

Swerdlow, Noel

[1977] "Pseudodoxia Copernicana: or, Enquiries into very many received tenets and commonly presumed truths, mostly concerning spheres," *Archives Internationales d'histoire des sciences*, vol. 26, 108-158.

Vernet, J.

[1976] "Jacob ben Machir ibn Tibbon," *Dictionary of Scientific Biography*, vol. 13, 400-401.

Zotenburg, H.

[1866] *Catalogues des Manuscrits Hébreux et Samaritains de la Bibliothèque Nationale*, (Paris).

IBN AL-HAYTHAM'S

ON THE CONFIGURATION OF THE WORLD

Chapter One

Introduction

In the Name of God, the Compassionate, the Merciful; Greatness is God's. The treatise of al-Ḥasan bin al- Ḥasan ibn al-Haytham on the Configuration of the World.

He said:

[1] Many of the mathematicians who have devoted careful study to the science of astronomy and attained the true facts concerning the forms found in the world as a whole and in most[1] of its parts have continued to expound an affirmed[2] doctrine in which they explain by way of a general theory how the world and every one of its primary parts[3] are pictured to be, the circumstances of all the heavenly bodies[4], their relative ordering, their distances from each other, the magnitude of their bodies, their various positions, the kinds of their motions and the varieties[5] of their shapes. However, in so doing, they proceed in a manner which conforms superficially to the detailed[6] investigation of the science of astronomy, drawing from those theories.[7] [2] Since those theories, that is, those which point to the form of the figure and the laws of the motions by means of proper observation and correct proofs are, however, based upon the motions of imaginary points on the circumferences of intellected circles according to what is demonstrated in those books of theirs which we have; and, likewise, they [those points] are assigned by indication on their part, but not explicitly, to the surfaces of solid spheres which, in fact, are the things which have those motions on those points[1], it turned out that their theory insofar as they explained it was limited to those circles and points only. [3] Their purpose was not to elucidate the manner by means of which those

various motions[1] may possibly be consummated, all the while that they are assumed [to take place] on the surfaces of solid spheres[2], nor[3] [did they explain] the particular circumstances of their deferent spheres with their various centers. They rather made it their objective to collect those details[4] of their circumstances which have been recorded and to present this in the form of remarks, devoid of proofs and [of a description] of the paths which led to their discovery. For their aim was none other than to make it easy for one who is interested in[5] acquiring knowledge of the configurations of those motions by way of acceptance, not investigation; by following the practitioners of the art, not through probing the understanding of that which has been set down concerning this; nor [questioning] the imagining of that which has been defined and described[6].

[4] Their doctrines, moreover, are not perfectly clear in all places, nor also are they exhaustive in all that the issues demand. Rather, they are persuasive in some places, gratuitous[1] in others. [5] We have not pursued in what we have said the path of calumny and rebuke, since it is possible that the authors of those treatises, though they did neglect the issues which they neglected, [nevertheless] conceived them and were versed in them. However, they did what was appropriate for the situation, and in that manner they cleared the way for theorizing about these issues due to their pioneering in their [the issues'] discussion.[1] They elucidated the difference between the theories in some aspects, though they [the theories] do agree in others.

[6] We are presenting[1] a doctrine on the configuration of the world which contains that which we hold concerning the totality of issues which the perception of the investigators has attained, according to the facts of the mathematical sciences and based on the goal of accurately conveying the sciences*[2] which the excellent Ptolemy demonstrated with the utmost scrutiny in his book on mathematics [the *Almagest*]. However, with regard to the motions, we shall assume that the imaginary circles and points which he adopted in the abstract [take their place] on the surfaces of spheres which themselves move with those motions, since this is more truly representative for the description of the situation and more clearly evident[3] to our understanding. [7] Our goal is to transmit that which we understand of these sciences in order to instruct him who wishes

---

* The Arabic *ᶜulûm* may also be taken to mean "items of knowledge."

to arrive at its comprehension without investigating. [8] He[1] [Ptolemy] was aiming to construct proofs for which it was sufficient to use those circles and points only.

[9] Since our doctrine is in accordance with what he [Ptolemy] explained and arranged, and he avoided[1] the use of any bodies, we investigated each of the motions which he mentioned in such a manner that that motion[2] may be appear to be the result of a spherical[3] body that is moving with a simple, continuous, and unceasing motion. [10] With this it is possible to join together all[1] those bodies which have been assumed for each one of the motions without there resulting[2] any hindrance, repelling, or impediment. Rather, their motions, including their combinations, are unceasing and continuous. [11] For each motion which we shall call "simple" we shall assume a spherical body moving about its center with a continuous motion since this is most appropriate for the eternal thing which is not subject to change and is free from defects. We shall then outline the doctrine concerning all[1] that which exists for the heavenly bodies[2] and the displacements and changes of position which befall them.

[12] Our doctrine concerning all of the motions[1] is, nevertheless, in accordance with the view of Ptolemy concerning them and with his conviction. All that which we present[2], however, are remarks by which the general form of the figure, the positioning, and the motion, [but] not the particulars of these, may be understood[3]. However, when their particulars are combined with them, they may [be brought to] agreement[4] with them. Our effort at this will be in accordance with our aptitude and ability.

Introductory note: Chapter one is missing from H6, H7, and L1.

Salutation

All the Hebrew manuscripts which have this chapter begin with *'amar* = Arabic *qâla* (he said), followed by the name of the author, which is given in most mss. as Abu ^{c}Alî bin [or ben, if we read this a Hebrew word] al- Ḥasan bin al-Haytham; H4, 5 read Abu ^{c}Alî (actually "al-^{c}Alî" in the manuscript) al- Ḥasan bin al-Ḥasan bin al-Haytham; H8 has Abu ^{c}Alî bin al-Haytham.

[1] 1. most: all (H2-4)

2. They explain . . . affirmed: H8 has translated *mursal* literally, i.e. "sent".

3. true facts . . . parts: the truth of the form which exists for the whole world and its most noble (*nikhbad*) parts (H8).

4. bodies: + *ha-ri'shonim* (primary) (H8).

5. varieties: *p^{e}raṭei* (details) (H8).

6. precise: - (H2-5).

7. drawing . . . theories: *l^{e}fi ha-nir'eh* (according to what is apparent) (H8).

[2] 1. which have . . . points: which move by themselves and with the motions of those points (?) (H8). The Hebrew is unclear.

[3] 1. motions: points (H8).

2. spheres: + *'aḥerim* (other) (H2, 4); + *'aḥadim* (some) (H5).

3. nor: *'ela hu* (but it is) (H8).

4. those circumstances...separately: *bi-hevdelei ᶜinyanam* (concerning the differences of their circumstances) (all H except H8), i.e., what is mentioned concerning their different circumstances, apparently taking the Arabic form of the root *f.ṣ.l.* in the sense of "separate", "different".

5. who is interested in: who thought that it would be useful (*she-yo ᶜil*) (H1, 4, 8). This is a perfectly acceptable translation of *yastafīd* .

6. not through questioning . . . describe: not by sensing (*murgashim*)(!) the understanding of that which they brought forth and the conception (*ṣiyyur*) of that which they defined (H8). The other Hebrew manuscripts are unclear here. The sense of these manuscripts seems to be that the treatises of earlier astronomers are not fully comprehensible. According to this interpretation, Ibn al-Haytham is beginning here the criticism which, in our reading, starts only in passage 4.

[4] 1. gratuitous: All H understand *mutasammiḥah* to mean "lacking" or "insufficient."

[5] 1. However, they did . . . discussion: but they took the gist (*to'ar*) of the matter and handled it in such a way so as

to be able to speak of these matters with (*ᶜim*) those who have previously spoken about them (H8).

[6] 1. presenting: *n ᵉ'ammet* (verifying) (H8).

2. based . . . science: and we rely on the intention (?) (*kawwanat*) of all the sciences (H8). H1 reads similarly, but has "the laws (*mishpatei*) of the sciences" instead of "all the sciences"; the other H read like H1, except that they read "and they relied."

3. and more clearly evident: Arabic Y reads here instead "and clearer also."

[8] 1. He: H8 seems to imply that the subject of this sentence is Ibn al-Haytham's treatise. Though this approach to this passage may be possible grammatically, it is most likely the incorrect interpretation, since there are no proofs in our text.

[9] 1. he avoided: it would be difficult (all H save H2): H2 reads, "it would be sufficient."

2. Since . . . motion: Because, given the structure and arrangement of our treatise, it would be difficult to use any body, we sought for each of the motions which we mentioned a way (*ṣad*) in which they may rotate and those motions . . . (H8) and so on as in our text. As with the previous variant from H8, it appears that H8 did not think that Ptolemy was the subject of Ibn al-Haytham's remarks. In this passage, H1 interjects "in my opinion," H2, "our opinion," and H3–5 the more correct "in his opinion."

3. spherical: particular (*juz'ī*) (Arabic Y).

[10] 1. all: - (Arabic K).

2. here resulting: *she-yaqsheh* (making it difficult) (all H except H8). Hindrance, etc., is the subject of this verb.

[11] 1. all: - (H2).

2. heavenly bodies: the bodies of the orbs (H8).

[12] 1. motions: positionings (Arabic K).

2. we present: he wished (H4, 5). Arabic K makes the subject "he" as well.

3. by which may be understood: $y^{e}khal\ ^{e}lem$ (which they comprise (H3-5).

4. agreement: + equally (H3-5).

Chapter Two

The Whole World

[13] The world is the name which is given to the collection of all existing things. The shape of the[1] world in its entirety is the shape of a sphere. The sphere is the form of a solid which is encompassed by a single surface in the middle of which there exists a point [such that] all the lines drawn from it to the encompassing surface[2] are equal. This point is the center of the sphere. [14] The world is a dense[1] body. Its shape is the shape of a sphere which is encompassed by a single surface in the middle of which there exists a point [such that] all lines drawn from it to the encompassing surface are equal. This point is called the center of the world.[2] I mean by saying "dense" that there is no empty space in it, but [rather] all of it is a plenum.[3]

[15] The bodies which fill the world are of different sorts, but they are on the whole divided into three divisions: heavy, light, and neither heavy nor light. [16] The heavy is that which moves from all regions of the world[1] to the center. The light is that which moves from the center to all regions of the periphery.[2] That which is neither heavy nor light is that which always moves around the center.[1]

[17] The special place of the heavy part of the world is the middle of the world. For this reason its motion is from all regions[1] of the world to the center so that it may settle in the place to which it belongs by nature. This heavy[2] part is divided into two categories, earth and water. [18] The earth in all its parts and that of it which is found in animals and plants is heavy by nature. Its proper place is the middle of the world, which is the closest place to the center. [19] If any part of it is raised[1] from its place to a place which is further away[2] from the center, and it is then released from it[3] [this place], it will move by the natural power which it has towards the center, and it will continue to do so unless an obstacle which comes between it[4] and the center impedes it. It then stops[5] since there is no path for the motion . [20] If the obstacle which is beneath it is removed, it then moves to the lower place. If it were possible[1] to do this forever, it would continue to move until it reached [the place]

where the center of the world would be in its middle.[2] So it is for every particle of water.

[21] The shape of the earth in its entirety and all its parts is like the shape of the sphere. However, its surface is not of perfect roundness; but, it has[1] corrugations which come about due to the effect of the motions[2] of the heavenly bodies.[3] However, this does not nullify its sphericity or change its shape. They [the corrugations] are rather in relation to[4] its whole, like the roughness which occurs in the surface[5] of some small[6] spheres. [22] The earth as a whole[1] is a round[2] sphere whose center is the center of the world. It is stationary in its [the world's] middle, fixed in it[3] and not moving in any direction nor moving with any of the varieties of motion, but rather always at rest.[4]

[23] Water[1] surrounds the sphere of the earth. However, since it [water] is heavy, and its motion is toward the center, and [since] the sphere of the earth stands between it and the center, it ends up in the place closest to the center[2] and surrounds the sphere of the earth. [24] Since the earth has an irregular exterior[1], having[2] in it low and high places; and since water, due to its heaviness, seeks the center and what is near the center,[3] it ends up descending by nature to the low places, and the high places remain uncovered, like the islands[4] in the middle of the sea, which the water surrounds. [25] The inhabited [portion] of the earth[1] is found in these islands[2] which jut out of the water[3]. [26] The exterior[1] surface of the water is the surface of a section of a sphere of perfect roundness and of equal[2] parts whose center is the center of the world. The lines drawn from the center of the world to all points[3] on the surface of the sphere of water are equal and likewise [for] every particle of water, even if separated like the pools and swamps.[4] This is the whole doctrine of the heavy parts.

[27] The special place of the light part of the world is in that which lies between the heavy part and the part which is neither heavy nor light. [28] For this reason, that of it which is held within the confines of the heavy part[1] will always move upwards when the obstacles are removed, until it reaches this place, i.e., that in which are the two remaining parts. The finer[2] part of it is arranged above the denser[3]. [29] This part, i.e., the light part, is divided into two

parts, air and fire. [30] Air[1] surrounds the sphere of water[2], and its shape is spherical. Its interior surface, which is adjacent[3] to the sphere of water, is a concave surface, spherically concave[4], because the surface[5] of the water bounds it, except that the high places of the earth dent it according to the corrugation of the surface of the sphere of the earth. However, this does not affect[6] its sphericity[7] because of the smallness of this amount [of indentation] compared to its totality.[8] Its upper surface is a convex surface which is adjacent[9] to the sphere of fire.

[31] This body, i.e., air, is of different parts and is subject to the accidents of generation and corruption[1]. Some of its parts are dense[2] and some are light.[3] [32] The density which is in it is due to the thick vapors which come about from the heating of wet bodies[1] as they are move upwards,[2] such as[3] the violent winds and the clouds[4] composed of wet vapor, or the coldness which comes about in wet air and the freezing of any of it, like the snow[5] which occurs,[6] and, in general, by the types of successive alteration. [33] That air which is adjacent[1] to the earth is warmer than that which is distant from it because of the reflection of the rays of the sun[2] which fall on the dense[3] body [of the earth]. The air becomes increasingly colder the more distant it is from the earth, until it comes near the sphere of fire. It then again becomes hot because of its fineness and[4] its nearness to the likeness of fire. Its likeness to it[5] [fire] intensifies the nearer it comes to fire, until it arrives at the body of fire[6] with which none of the air can mix.

[34] Fire surrounds the sphere of air[1], and its shape is a spherical[2]one. Its interior surface is spherically concave, surrounding the sphere of air[3]. Its exterior surface is a convex spherical surface of perfect roundness; the body surrounding it and moving on it bounds[4] it. This is the whole doctrine of the two light bodies.[5]

[35] The part which is neither heavy nor light surrounds the two remaining parts and moves around them with an unceasing motion. This part is that which is called the orb, and it is that which contains[1] all the stars. [36] Its shape in its entirety[1] is a spherical shape. It is surrounded and bounded by two spherical surfaces[2], whose center is the center of the world. One of them, the interior[3], is

concave, of perfect roundness[4], and tangent to the sphere of fire. Its center is the center of the world.[5] The other, outer one, is a convex spherical surface of perfect roundness and[6] of similar parts, whose center is[7] the center of the world.[8] It surrounds all existing things.[9] [37] This body in its entirety moves with a swift motion from the direction that is called east to the direction that is called west.[1] It sets in motion by means of its own motion all the heavenly bodies[2] which are contained within it with a circular motion. [38] This body is divided into parts.[1] We will explain each one of them in its place. This is the summarized doctrine of the whole universe, and we will complete its elaboration, God—may He be exalted—willing.[2]

[13] 1. The shape of - (H2).

2. plane: sphere (H3-5).

[14] 1. dense: - (L1).

2. This point . . . world: - (H8).

3. I mean . . . plenum: - (L1).

[16] 1. of the world: - (all H except 6, 8).

2. periphery: surface (H4, 5); world (H6).

[17] 1. all regions: *me-* ^{c}*elyonot* (from high places) (?) (H6).

2. heavy: - (L1).

[19] 1. raised: removed (all H, though different words are used).

2. further away: + *superiori* (L1).

3. and it is then released from it: Only H8 translates the Arabic accurately. H6 may be translated, not without difficulty, "and it is left (*we-yanu'aḥ*) to its nature"; the other H read, "when the removing agent (*ha-mesir*) goes away."

4. which comes between it: This phrase offers an interesting example of the independence shown by the Hebrew manuscripts. Four different terms are found here, though all mean essentially the same thing as the Arabic. H1, 2 have *ya* ^{c}*amod* , to stand; H3, *yafrid* , to separate, come between; H4, 5 *yasdir* , to arrange [itself, i.e. to become placed]; and H6, 8, *yavdil* , to differentiate, separate.

5. stops: H1, 2 read like the Arabic, as do H4, 5, though the latter have the same spelling mistake, reading *yit* ^{c}*aqqev* instead of the correct *yit* ^{c}*akkev* . By changing one letter, H3 reads *yit* ^{c}*arev* , "to mix." H6 employs two synonyms:

ya ᶜamod ve-yishqoṭ . This is not an unusual practice for Hebrew translators in cases of uncertainty.

6. If it were possible: *Et si posibilie esse quod nullum invenisset instans* (L1).

7. If it were . . . middle: - (H4, 5).

[21] 1. it has: it is mostly (H1, 2).

2. the motions: This is missing in Arabic Y but found in all the translations.

3. the heavenly bodies: the orb (H8).

4. in relation to: *bi-hiṣṭaref* (in addition to) (H1). This is equally acceptable as a translation of the Arabic .

5. surface: - (H8).

6. small: - (H3).

[22] 1. as a whole: - (L1).

2. round: - (H1, 6).

3. fixed in it: - (H6).

4. but rather always at rest: - (H6).

[23] 1. Water: - (H8); + the body of (H6).

2. the place . . . center: outside of the center, close to it (H6).

[24] 1. has an irregular exterior: There are several quite different renderings of this in the Hebrew. H1 has: is smooth (*ḥalaqah*) in appearance; H4, 5 change *l ᵉfi ha-nir'eh* (in appearance) to *l ᵉfi kavdutah* (due to its heaviness); H6 reads: the parts of the earth break apart (*mitrapp ᵉqim*) one from another. H8 gives the reading which is closest to that of the Arabic: is unequal in appearance.

2. an irregular exterior, having: - (L1).

3. and what . . . center: - (H6).

[25] 1. earth: + surface of (H6).

2. islands: rocks (H6).

3. which jut . . . water: - (L1).

[26] 1. exterior: - (L1).

2. equal: *consimilium* (L1) = *mitdammeh* (all H).

3. points: places (H8).

4. The lines . . . swamps: - (L1). swamps: rivers (all H - 8).

[28] 1. which is . . . heavy part: *quod clauditur ex ea in in firmatibus infra partem ponderosam* (L1).

2. finer: - (H1, 2)

3. this place . . . heavier: its place from which it is thicker (H6); heavier: - (H2-5).

[30] 1. Air: + fire (H3-5).

2. water: *terre* (L1).

3. adjacent: opposite (*neged*) (H8).

4. surface, spherically concave: - (L1).

5. surface: - (H2-5).

6. does not affect: is not noticeable (H1-5); does not nullify (H6, 8); affect: *destruens* (L1).

7. sphericity: *denticitatem spericitatem* (L1).

8. its totality: its total body (H6); the total body of the air (H8).

9. adjacent: tangent (all H - 6); *breviatur* (L1).

[31] 1. accidents . . . corruption: *immutationem que accidit in ea continue* (L1).

2. dense: H6 adds here *w^{e}-gasim* , a synonym for the term *^{c}av* (dense).

3. light: Again, H6 adds a synonym, *rafim* .

[32] 1. bodies: + earthly (H6, 8).

2. as...upwards: - (H6). H4, 5 are missing "upwards."

3. such as: *et propter id quod generatur ab eis* (L1).

4. clouds: + dense (H6).

5. or the coldness like the snow: into the snow (all H); snow: + hail(H6); snow: and its freezing into snow (H8).

[33] 1. adjacent: tangent (all H- 8); H8 reads "near."

2. sun: + and the stars (H2).

3. dense: hard (H6).

4. its fineness and : - (H8, L1).

5. likeness to it: heat (H1-3).

6. until . . . fire: *usque quo transeat in naturam ignis* (L1).

[34] 1. air: fire (H4,5).

2. spherical: - (H8).

3. and its shape . . . air: - (H3-5).

4. bounds: like Arabic Y, all the Hebrew manuscripts except H6 add here as a synonym *w^{e}-y^{e}'amm^{e}ṣehu* [literally: to strengthen, to firm up], except H2, which has*w^{e}–y^{e} ḥofifehu* .

5. is a convex . . . bodies: is on top of the two light parts (H8).

[35] 1. contains: + and encompasses (H2).

[36] 1. in its entirety: - (L1).

2. surfaces: + *equidistantes* (L1).

3. the interior: - (H1).

4. of perfect roundness: Here H7 begins; roundness: + *consimilium partium* (L1).

5. and tangent . . . world: - (L1).

6. and tangent . . . roundness and: - (H6).

7. is: is not (!) (H2).

8. The other . . . world: - (H8); and of similar . . . world: - (L1).

9. things: + corporeal (H2).

[37] 1. from the direction . . . west:*ab oriente in occidentem* (L1).

2. heavenly bodies: the bodies of the orbs (H8).

[38] 1. parts: + *multas* (L1).

2. This is . . . willing: *cum Deo auxilio* (L1).

Chapter Three

The Orb

[39] Orb is a designated term* which is said of any round magnitude of a spherical body,[1] a spherical surface,[2] a plane of a circle, or the circumference of a circle. The part[3] surrounding the world and moving around the center is called in its entirety an orb. [40] This orb is divided into many parts. It is, however, primarily divided into nine parts, which are contiguous[1] spherical bodies, one of them surrounding the one adjacent to it,[2] the concave[3] surface of the surrounding [orb] touching the convex surface of the one surrounded by it.[4] The center of all these spheres is the center of the world.

[41] Each one of these individually is [also] called an orb. [42] The first of them is the orb of the moon. Next to it is the orb of Mercury, and next to that[1] is the orb of Venus. Then, next to it is the orb of the sun,[2] then the orb of Mars, then the orb of Jupiter, then the orb of Saturn, then the orb of the fixed stars, then the surrounding[3] greatest orb which sets in motion all these orbs. [43] These orbs have different motions, but in general[1] they are divided into two opposite motions,[2] one of which is from east to west, the other from west to east.[3]

[44] The greatest orb which surrounds all the orbs moves with a swift motion from east to west[1] on two fixed poles. It makes one revolution in every day and night. All the orbs which it surrounds move with its motion. [45] One of the two fixed poles is called the north pole because it is near the region which is called[1] north.[2] This region is the inhabited region of the earth, and this pole is the one visible to the people living in the north. [46] The other pole is called the south pole, because it is near the region which is called south, and this pole is hidden[1] from the people living in the north.

[47] As for the remaining orbs, some of them move from west to east[1], and some also move contrary to that [direction]. Every one of them has a motion on the whole[2] from west to east or from east to

---

* The Arabic phrase is *'ism mawḍû* c ,literally "a name that has been set down."

west. However, they [the orbs] are on different poles. Each one has two poles which are specifically associated with it.[3] [48] Each one of these spheres is divided into different[1] spheres,[2] each one of which has a motion specifically associated with it. We will speak about each one of them in its place, God willing.[3]

[49] Every sphere[1] moves on two poles. The points on it trace circles on the surface of the sphere; this has been demonstrated by the geometers in their books. Every plane which intersects contiguous spheres describes[2] a circle in each one of them; this has also been demonstrated. [50] When these spheres move on their poles, each star in them describes an imaginary circle in the sphere by means of the point of its [the star's] center. If we imagine the plane of this circle to intersect the world* ,[1] it will then produce a circle in the surface[2] of the highest[3] orb also. Similarly, if we imagine a plane passing through the center of the star and the center of the world,[4] it will produce a great circle whose center is the center of the world[5] on the surface of the highest[6] orb.[7] Many imaginary circles[8] of these two sorts are produced in the surface of the highest orb. Each one of them has a special form.[9]

[51] When the greatest orb moves with its swift motion, all the points on it describe circles[1] on the surface of the greatest orb, as we said before. All these circles are parallel to each other.[2] Their poles are the poles of the world,[3] and they are of different magnitude. The closer they are to the poles, the smaller they are, and the further they are from the poles, the greater they are. The greatest[4] of these circles is the one which passes through the center of the world. It is equidistant from the two poles. It divides the world in half because it passes through its center. [52] This circle is called the celestial equator** because, when the sun reaches it,[1] night and day are equal in the whole world.[2] The poles of this circle are the poles of the world which we called the north and south[3] [poles]. [53] That which we call the pole of a circle[1] is the point from which all lines drawn to

* That is, if we imagine it to intersect all of the celestial orbs.

** "celestial equator" : literally, "the circle which makes the daylight equal."

the circumference of that circle are equal. The remaining parallel circles are called time circles*.

[54] These circles and all the circles in the orb are divided into 360 equal parts by means of an imaginary division. Each part is called a degree. These parts of the time[1] circles are called the parts of time. [55] These circles are also divided into 24 divisions, each of which is 15° and is called an hour. The whole circle is 24 hours.

[56] When the greatest orb turns, it sets all the stars in motion. Each of the stars moves on one of these circles. Then, whenever a point[1] on the orb moves through 15°, it is said that one hour has passed. When it has moved through the whole circle the elapsed[2] time is 24 hours. [57] It [the point] moves through the circle in one revolution[1] which is imparted to it by the greatest orb. Day and night occur with the complete revolution of this orb. For this reason, day and night are 24 hours.[2]

[58] We imagine also a great circle which passes through the center of the world and the two poles. It cuts the circle of the celestial equator in half, it also intersects the world in its entirety, and it intersects the sphere of the earth. [59] The circle of the celestial equator also cuts the sphere of the earth in half. The earth is divided by these two circles into four equal parts, two parts of which are to the north, and two parts of which are to the south. [60] One of the two northern parts is the inhabited quarter of the earth. The circle which is produced on the surface of the earth by the intersection of the celestial equator is called the terrestrial equator. The half[1] of this circle which bounds the inhabited [part] of the earth is the "length of civilization" and it is 180 parts[2] of the parts of the circle. [61] In the middle of this length, on the terrestrial equator** , is a place[1] which is called the middle of the earth, and it is called the "dome."*** [3]

---

* This is the literal translation of *dawâ'ir al-azmân* . These are the parallels of declination, along which the star's hour angle may be measured.

** "terrestrial equator" : literally, "the line of equality."

*** Concerning this dome (Arabic *qubbah*) see ᶜAbd al-Raḥmân Ḥamîdah, *Aᶜlâm al-Jughrâfiyîn al-ᶜArab wa-Muqtafât min Athârihim* (Damascus,1969), pp. 92-93; and Carl Schoy, "The Geography of the Moslems of the Middle Ages," *Geographical Review* 14(1924)257-269, pp. 263-64.

[62] The other circle passing through the two poles and dividing the earth is called the horizon of the middle of the earth and is [also] called the horizon of the terrestrial equator.[1] This circle divides the world in half.[2] One of the halves[3] is visible to someone who is at the middle of the earth, and the other half[4] is hidden from him. [63] Every great circle which divides the world in half and which forms a boundary between what is visible and what is not is called an horizon. The place which is in the middle of the earth is like the pole of the circle which passes through the two poles[1] [of the world] and forms the boundary of the inhabited world. [64] When the line which is drawn from the center of the earth to this pole is extended rectilinearly until it reaches the highest orb, the point at which it terminates on the surface of the highest orb is called the zenith because it faces the heads of those who are[1] in the middle of earth. That point is in truth a pole of the great circle which is the horizon of the middle of the earth. [65] For every point on the earth, the point facing it on the highest orb is called the zenith of that place on the earth. The great circle of which this point is the pole is the horizon circle for that place on earth.[1]

[66] For every point on the earth there is an horizon which divides the world in half. [67] The two poles of the world are on the circumferences of the horizons of the points which are on the terrestrial equator.[1] The circle of the celestial equator is equally perpendicular to each one of them. [68] One of its [celestial equator's] poles is visible in the rest of the horizons and the other is hidden. The celestial equator is inclined to them [the remaining horizons]. The circle of the celestial equator is inclined from the horizons of those living in the north[1] towards the south, and in them [these zones] the north pole is visible above the earth.[2] As for the southern horizons, the celestial equator is inclined in them towards the north, and the southern pole is visible in them above the earth.[3]

[69] The great[1] circle which we imagine to pass through the poles of the world and the zenith point in any[2] of the horizons is called the meridian circle* for that horizon because[3] [of the fact that] when the sun reaches it in its swift,[4] daily motion, daylight is half over in that place.[5] Likewise, when it [the sun] reaches it in the half

* "meridian circle": literally, "circle of half daylight."

which is below the horizon, night is half over.[6] [70] This circle divides the horizon in half. Two points result from their intersection. One of them is to the north and is called the northern point; the other is to the south and is called the southern point. [71] Similarly, the celestial equator cuts the horizon in half. Two other points result. One is called the eastern point and the other the western point.

[72] In every horizon there are four points, namely north, south, east, and west.[1] The straight line which runs between the eastern and eastern points is called the east-west line.[2] [73] The straight line which runs between the northern and southern points is called the north-south line. This line is also called the meridian line* . It is that line which is drawn on the surfaces of sundials. It is the intersection of the meridian circle and the plane of the horizon circle.[1]

[74] Again, we imagine a great circle to pass through the zenith, which is the pole of the horizon, and the center of the world. It is perpendicular to the horizon, intersecting its circumference in two points.[1] It intersects the plane of the horizon in a straight line[2] which runs between these two points and it passes through the center of the world. This circle is called the zenith circle.[3] [75] Similarly, all the circles which fit this description are called zenith circles. Each of these two points is called the zenith point.[1] The straight line is called the zenith line.[2] [76] The arc of the horizon which lies between the northern or southern point and the zenith point is called the zenith arc. Sometimes the arc between the eastern or western point and the zenith is called a zenith [arc].[1]

---

* "meridian line" : literally, "line of half-daylight."

[39] 1. body: - (H8).

2. a spherical surface: - (all H except H6, 8).

3. the part: the body (H6, 8).

[40] 1. contiguous: - (H8).

2. one of them surrounding . . . to it: *scilicet quod* (L1).

3. concave: - (L1).

4. convex surface . . . surrounded by it: *superficiem connexam* (L1).

[42] 1. next to that: and on top of (*ᶜal*) the orb of Mercury (H8).

2. next to it is the orb of the sun: - (H4, 5)

3. surrounding: - (all H except H6, 8).

[43] 1. in general: first of all (H6).

2. but in general . . . motions: - (H8).

3. These orbs . . . from west to east: - (L1).

[44] 1. The greatest . . . east to west: - (L1).

[45] 1. which is called: - (L1).

2. north: + *partes* (L1).

[46] 1. hidden: + *ad aspectu* (L1).

[47] 1. some . . . east: - (H4, 5, 8).

2. on the whole: - (L1).

3. Each one . . . with it: - (H8).

[48] 1. different: *duas* (L1).

2. Each one . . . spheres: - (H8).

3. God willing: - (L1). This is missing from Arabic Y.

[49] 1. sphere: motion (H2, 3, 5 7).

2. Every plane . . . describes: - (H8).

[50] 1. to intersect the world: to pass through the circumference and center of the world (H8).

2. surface: - (H6).

3. highest: the great (H2-5, 7)

4. world: + *secantem ipsum* (L1).

5. great . . . world: a circle (H8). Some lines are repeated in the manuscript.

6. highest: ecliptic (H6).

7. Similarly . . . orb: H2-5, 7 combine parts of this sentence with the previous sentence. They read: If we imagine the plane of this circle to intersect the world, there results a great circle whose center is the center of the world.

8. Many imaginary circles: *circuli principales ymaginari* (L1).

9. Each one . . . form: - (L1). form: *ᶜinyan* (H6). This word is usually used to translate the Arabic *ḥâl* . The meaning thus becomes: Each one is a special case. H8 adds *sᵉgulah* , which has the same effect on the meaning.

[51] 1. circles: + *equidistantes* (L1).

2. All . . . each other: - (L1).

3. of the world: *ipsiius* (L1), referring to the highest orb.

4. greatest: furthest (H6).

[52] 1. reaches it: that day that it moves on it (H8).

2. world: + *praeter quam in duobis locis in quibus habitantium capite directe respiciunt duos polos* (L1); there is a similar addition in H6, 8 (H8 adds: the poles of the world). This addition is important because it agrees with the addition found in Arabic K, which in turn quotes this phrase as belonging to another version of our text.

3. which we . . . south: - (L1).

[53] 1. circle: + *b e-kadur* (in a sphere) (H6); similarly, L1 adds: *in spera est punctus in superficie illius spers* .

[54] 1. time: other (H4, 5).

[56] 1. point: star (H6); part (H8).

2. elapsed: - (H4 - 7; L1).

3. It moves . . hours: - (L1).

[57] 1. revolution: + complete (H6).

2. It moves . . . hours: - (L1). All the Hebrew manuscripts add: approximately, more literally "according to imprecise thinking" (*ciyyun ha-gas*).

[60] 1. half: part (H2).

2. parts: *ma calot* (degrees) (H6). Note that here H6 uses the term used in H1-5, 7, instead of *madregah* , the word usually used by Ibn Paṭer .

[61] 1. place: *linea* (L1).

[62] 1. and . . . equator: - (H2-5, 7).

2. in half: - (L1).

3. of the halves: - (L1).

4. half: - (L1).

[63] 1. poles: circles (H2-7).

[64] 1. who are: *ha-shokhnim* (who inhabit) (all H).

[65] 1. The great . . . earth: - (Arabic Y).

[67] 1. The two . . . equator: - (L1).

[68] 1. The circle . . . north: - (H4, 5, 8); north: south (!)(H8).

2. earth: *orizontem* (L1) = (H6, 8).

3. above the earth: - (H8). As before, H6 and L1 read here "horizon."

[69] 1. The great: This (H2).

2. any: all (H6).

3. because: Arabic Y adds here, "and it is called the meridian circle because...". This addition is found in all the Hebrew manuscripts except H4-6.

4. swift: - (H6).

5. in that place: - (L1).

6. night is half over: - (H8).

[72] 1. namely . . . west: *predicti* (L1).

2. The straight . . . east-west line: - (all H except H1). This sentence is also missing from Arabic Y. Its presence in H1 is one of the indications that H1 represents an independent effort.

[73] 1. common section . . . circle: *et ipsa est directa communis (?) circulo meridiei orizonti* (*sic* in L1); meridian... circle: the circle of the meridian plane and the horizon circle (*sic* in H6).

[74] 1. intersecting . . . points: - (H6).

2. line: plane (!) (H2-5, 7).

3. zenith circle: H8 here gives two terms: *ᶜiggulah nᵉkhoḥit* , the term used exclusively in H1-5, 7, and *ᶜiggulat ha-qodqod* , which is used by H6, 8. In other words, the copyist of the Ibn Paṭer translation has decided to insert as well the presumably better known Tibbonite technical vocabulary.

[75] 1. zenith point: + eastern or western or southern or northern (H6).

2. zenith line: + The straight line which is the common section of the zenith circle and the plane of the horizon is called the zenith arc (H8). This same sentence is found in H6, where it replaces the next sentence of our text (which is present in H8).

[76] 1. zenith: + *arcus* (L1) = (H6, 8).

Chapter Four

Longitudes and Latitudes

[77] The latitude of a city is the arc of the meridian circle between the zenith and the circle of the celestial equator. [78] Facing it [this arc of the meridian] on the surface of the earth[1] there is an arc of a great circle which surrounds the sphere of the earth and whose center is the center of the earth[2], lying between any point you take in that city and the terrestrial equator, which [in turn] is determined[3] by the plane of the celestial equator[4]. [79] The meridian line which is drawn in the plane of the horizon of that city is in the plane of this circle[1]. This arc is the distance of the city from the terrestrial equator.

[80] The longitude of a city is an arc of the circle of the celestial equator lying between the meridian of that city and the end of the inhabited world[1], [drawn] from either the east or the west. This arc is a section of the semicircle of the celestial equator which, as we mentioned, forms the boundary of the inhabited world.[2] [81] The extremities of the inhabited world are the two points of intersection of this half[1] [circle] with the circle of the horizon [2], [i.e.] the horizon of the middle[3] of the earth.

[82] Some of the time circles, which are parallel to the celestial equator, are imagined to intersect the world and to intersect the horizons of the [various] places. Each of these circles intersects each horizon in two points, and the horizon [in turn] divides them into two sections. The horizons of the equator cut all these circles in half, because these horizons pass through the poles of these circles.[1] [83] The remaining horizons divide only the circle of the celestial equator in half, dividing these[1] circles into unequal sections. The sections that are above the horizon are not equal to the sections that are below the horizon. As one goes further from the celestial equator in the direction of the visible[2] pole, the section[3] which is above the horizon[4] becomes greater[5] than the corresponding one[6] which is close to the celestial equator. As one goes further from the celestial equator in the direction of the hidden [pole], the section[7] which is

above the horizon[8] becomes smaller than the corresponding one which is closer to the celestial equator.[9]

[84] Night and day vary with the differences of these sections in the inclined horizons. These arcs[1] are those which are called the arcs[2] of daylight[3]. For every star which moves on one of these circles[4], the higher arc of these circles is called the arc of daylight[5] for that star. Its lower arc is called the arc of night[6]. [85] When the sun moves on one of these circles, the parts of the higher arc of this circle are the parts of the hours of daylight. The parts of the lower arc are the parts of[1] the hours of night.[2] [86] As these arcs[1] get further away from the celestial equator[2] in the direction of the visible pole, that [part] of [each one of] them which is visible becomes "more similar" than that [visible part of a day circle] which is close[3] [to the equator]. I mean by "more similar" that whose parts by which the circle is divided into 360 parts is greater in number.[4] It is thus necessary that the parts of those arcs which are distant be greater in number. Therefore, since the parts of those arcs which are distant from the celestial equator are greater in number,[5] the sun's sojourn above the earth is greater when it moves on these more distant circles,[6] and the parts of the hours of daylight are greater.[7]

[87] Of the circles which are near the north pole, the last one on which the sun moves is called the circle of Cancer.[1] Daylight is longest in the northern horizons when the sun moves on this circle. [88] The contrary to that necessarily follows with regard to all the circles which are near the south pole. That is, the more distant the arc which is above the horizon[1] becomes from the celestial equator, the "less similar"[2] it becomes than that which is close. Likewise, as the sun becomes more distant[3] to the south, daylight becomes shorter. The last of these circles[4] on which the sun moves[5] is called the circle of Capricorn.

[89] None of this[1] happens in the horizons of the terrestrial equator. Rather, the arc of daylight for all stars is equal to the arc of night, because the horizons divide all the circles parallel to the celestial equator[2] in half, and so the arc which is above the horizon is equal to the arc which is below the horizon.[3] For this reason, night and day are always equal in these places, because the sojourn of the sun above the horizon[4] is equal to its sojourn below the horizon.[4]

[90] Each of the parallel circles which intersects the horizons intersects them in two points. One [of the two points] is to the east and other is to the west. [91] Every star which moves on that circle[1] rises from the rising point and sets at the setting point opposite it.[2] Similarly, every point[3] on the surface of[4] the highest[5] orb rises from one of these points and sets at the setting[6] point opposite it. [92] The arc of the horizon which lies between the rising point and the eastern point[1] of the celestial equator is called the rising amplitude for that star or for that point which rises from the rising point of the horizon.[2] The arc of the horizon which lies between the setting point and these western point of the celestial equator is called the setting amplitude* .[1]

[93] We have already mentioned that the closer the parallel circles are to the visible pole, the bigger the arc which is above the horizon. It continues so until one reaches a circle which touches the horizon in one point. Then the whole circle[1] is visible above the horizon. [94] Every star which moves on this circle is always visible.[1] The contrary follows with regard to that which is near the hidden pole, until one reaches a circle which touches the horizon in one point. Then all of it is below the horizon,[2] and the star which moves on it is always hidden.

[95] [Moving] beyond these two circles[1] : all of those which are near the visible pole are visible [in their entirety], elevated above the horizon; and all those which are near the hidden pole are hidden, lowered[2] below the horizon.[3]

[96] The entire[1] inhabited (part) of the earth, according to what was said,[2] lies in that which is between two parallel circles. One of them is the circle of the celestial equator, which passes through the zeniths of those who are on the terrestrial equator. The other is the circle whose distance from the celestial equator is 66 parts of the parts of which 360 comprise the circle. This circle passes through the zeniths of those who are at the end of the inhabited world. In that place the circle of Cancer[3] is tangent to the horizon.

---

* The rising amplitude is also called the ortive amplitude, and the setting amplitude is also called the occassive amplitude. For the calculation of these quantities see O. Neugebauer, *A History of Mathematical Astronomy* (Berlin-Heidelberg-New York), vol. 1, pp. 37-39.

[97] This section of the earth is divided into seven divisions which are called seven climates.[1] [98] The beginning of the[1] first of the[2] climates is the terrestrial equator. Night and day are always[3] equal in this place. Then night and day continue to differ [with respect to their length] for all those who are close to the north, as we said before, until one reaches the end of the[4] seventh climate, which is the end of the inhabited world. [99] In that place the longest daylight is 24 hours. This is because the circle of Cancer is tangent to their horizon in one point, and hence all of it is visible above their horizon. Then, when the sun reaches this circle, it is above the horizon for its full revolution.[1] For this reason, their[2] longest daylight is 24 hours. Then the sun returns[3] [to the south], and daylight continues to decrease and night to increase, until the sun reaches its furthest distance in the south. When it [the sun] reaches[4] the circle of Capricorn, it is below[5] the horizon for its full revolution[6], because the circle of Capricorn is tangent to the horizon in one point, and all of it is below the horizon.[7] Thus their[8] longest[9] night is 24 hours. [100] The latitude of this place, which is the arc of the meridian circle between the zenith and the celestial equator, is close to 66 parts of the parts of which 360 comprise the circle.[1] It is said that there is nothing of the inhabited world in that which is beyond this circle* .

[101] The seven climates have been divided in another way. There are those** who make the beginning of the inhabited world

---

* For a medieval Islamic report of life in the far north, see al-Bîrûnî's *Kitâb Taḥdîd al-Amâkin* (ed. Bulgakov), pp. 137-138.

** It seems to me that Ibn al-Haytham has in mind here the theory which recognized that human civilization extended further to the south than the limit fixed by Ptolemy. The debate on this point was summarized by al-Mas^c ûdî (*Al-Tanbîh wa-l-Ashrâf* , Baghdad,1938, p. 31): "Those scholars and philosophers of the nations who are concerned with the science of geometry and the measurement of the earth contend with one another concerning the seven divisions, [viz.] whether they are in [both] the north and the south or in the north but not in the south. The majority hold that they are in the north but not in the south due to the greatness of the population in the north and its sparsity in the south. People are of the opinion that the ancients intended the division of the seven climata for the northern side of the equator, and they made no division on the southern [side] because of the small amount of population to the south of it. Hermes held...that there are seven climata in the south just as there are in the north." This ancient controversy took on new

other than the terrestrial equator and its ending other than[1] latitude 66.[2] That which lies outside of that[3] [area bounded by the two circles] is not considered to be part of the inhabited world.[4] [102] Ptolemy made the end of the inhabited world in the north latitude[1] 63,[2] and he made its end in the south[3] beyond[4] the terrestrial equator at latitude 16 + (1/4) + (1/6)[5], according to what he said in his book on the inhabited world.

[103] As for that[1] which lies beyond latitude[2] 66 of the earth: as one gets closer to the north[3], the circle of the circle of Cancer gets higher above the horizon. Then the horizon ends up tangent to a circle greater than it [the circle of Cancer]. [104] As long as the sun remains[1] on one of the parallel[2] circles which lie between the tangent circle[3] and the circle of Cancer, there will be daylight in that place, because the sun will be visible above the earth.[4] Such will be [the situation with regards to] night when the sun is[5] near the south pole.

[105] Night and day continue to increase the closer the place is to the visible pole[1], until one reaches the place which faces[2] the pole[3], i.e., that [place] at which the pole of the world is at the zenith and the axis[4] of the world is perpendicular[5]. The horizon in that case is the circle of the celestial equator. [106] The solar year in this place is a day and night[1]. Daylight[2] is close to six months [long], that is as long as the sun is in the northern signs, because it is [then] always above the horizon. Night is also close to[3] six months [long], that is, as long as the sun is in[4] the southern[5] signs, because it is [then] always below the horizon[6]. In this place the sun always moves around the horizon, not rising very high. It rises highest when it is in the circle of Cancer.[7]

---

urgency as the Arabs voyaged to Africa, India, and Southeast Asia, discovering human civilization in areas which had been supposed to be uninhabitable. See E. Honigman, *Die sieben Klimata* (Heidelberg,1929), p. 162, and S. M. Ahmad, "Djughrâfiyâ," *Encyclopaedia of Islam* , second edition, vol. 2 (Leyden,1965) 575-587, p. 578.

[78] 1. Facing...earth: *intersecat eum superficies orizontes* (L1). For "earth", H6,8 read "horizon".

2. earth: world (H6).

3. determined: described (*rasham*) (H6).

4. of the celestial equator: *huius circuli* (L1).

[79] 1. of that circle...circle: - (L1)

[80] 1. end...world: betwen the meridian circle and the city of the end of the inhabited world (*m ᵉdinat qaṣeh ha-yishshuv*)(H6).

2. world: + and the extremities of the inhabited world (H6).

[81] 1. of this half: - (H2-5,7). All H are inexact here except for H6,8.

2. the circle of the horizon: - (L1).

3. middle: *ḥeṣi* (half) (H8).

[82] 1. in half...circles: - (H4,5).

[83] 1. these: the remaining (H6).

2. visible: - (H8) = the aforementioned (H3,7)= the hidden (!) (H2,4,5).

3. section: arcs (all H except H6).

4. horizon: earth (H2-5,7).

5. greater: smaller (!) (H2-5,7).

6. one: + of that circle (H6,8).

7. section: arcs (H1).

8. horizon: earth (H1). This is also the reading of Arabic Y.

9. The last two sentences of this passage are missing from L1. The last sentence is missing also from H2-5,7.

[84] 1. arcs: sections (Arabic Y).

2. arcs: sections (Arabic Y).

3. arcs of daylight: *archus diei et noctis* (L1).

4. one of these circles: + which are parallel to the circle of the celestial equator (H6) ; H8 adds instead, "i.e., that the position of the star on the surface of the orb moves on circles parallel to the celestial equator."

5. of daylight: - (H8).

6. night: + for that star (H6,8).

[85] 1. the parts of: - (H1,2;L1)

2. The parts...night: - (H3).

[86] 1. these arcs: *hii circuli magis* (L1).

2. As...equator: - (H3)

3. than that which is close: *quod apparet ex circulo magis appropinquare* (L1).

4. greater in number: *principales* (L1).

5. the parts...number: - (L1).

6. circles: arcs (H6).

7. and the parts...greater: - (H8,L1). H8 does add here, "from the celestial equator."

[87] 1. Cancer: *ḥaṣi sarṭan* (half Cancer)(!) (H4,5).

[88] 1. which is above the horizon: *apparentes* (L1).

2. "less similar": smaller than the corresponding one (H8).

3. becomes more distant: is seen (H6).

4. of these circles: motion (H1).

5. on which the sun moves: - (H6).

[89] 1. this: + *shinnuy* (variation) (H8).

2. Rather...equator: all the circles parallel to the celestial equator are divided (H3-5,7).

3. and so...horizon: - (L1).

4. horizon: *terram* (L1). This is the reading of Arabic Y.

[91] 1. which...circle: -(H6).

2. opposite it: - (all translations).

3. point: + *ymaginabili* (L1).

4. the suface of: - (H6).

5. highest: - (all H).

6. setting: - (Arabic Y).

[92] 1. eastern point: that which faces the eastern point (H6).

2. of the horizon: *qui est in oriente* (L1).

[93] 1. the whole circle: the circles (H1).

[94] 1. visible: + on the horizon (H8).

2. in one point...horizon: - (H1).

[95] 1. these two circles: - (H8).

2. visible: - (L1).

3. All Hebrew manuscripts except H6 read like our text. H6 follows the version of Arabic K, replacing "all the circles are hidden" by "the circles are always hidden" and similarly adjusting the rest of this passage. L1 combines the two readings.

[96] 1. entire: most of (all H save H6,8).

2. was said: Ptolemy said (H8).

3. circle of Cancer: circles of Cancer and Capricorn (H6,8).

[97] 1. which are...climates: - (all H save H6,8). H6,8 are lacking only the word "seven."

[98] 1. beginning of the: - (L1).

2. the first of the: these (H8).

3. always: - (L1).

4. the end of: - (H2-6).

[99] 1. Then...revolution: - (H6); it is...revolution: the whole circle is above the horizon (all H save H6).

2. their: *in hoc loco* (L1).

3. returns: *retrogradat* (L1). The Arabic *raja* ^{c}*a* means both "to return" and "to retrograde"; the sun, of course, does not display any retrograde motion.

4. When it reaches: *que est* (L1).

6. full revolution: the whole circle [is below...] (all H save H6).

7. because...below the horizon: - (H6); and all...horizon: This is missing from all the Hebrew manuscripts except H1.

8. their: *in hoc loco* (L1).

9. their longest: - (H6).

[100] 1. parts...circle: *madregot* (degrees)(H6).

[101] 1. below: other than (*zulat*) (all H save H6,8). H8 exhibits, "their beginning preceding (*qodem*) the equator and their ending less than latitude 66."

2. There are...66: *sic scilicet ut principium et finis eorum ponatur minus quam 66 gradum* (L1); cf. reading from H8 in previous note.

3. that: + *extremarum* (L1).

[102] 1. latitude: - (L1).

2. 63: 66 (H2,4,5) = 66 44 (H6).

3. south: other (H2,3,7).

4. beyond: *equidistans* (L1).

5. 16 + 1/4 + 1/6: Arabic K and L1 read "26" instead of "16", and the fraction 1/6 is not found in H6.

[103] 1. that: the circle (H8).

2. latitude: - (H4,5).

3. to the north: - (H3,7).

[104] 1. remains: moves (H6).

2. parallel: - (H7).

3. tangent circle: *contingens orizontem* (L1); similarly, H6 adds, "to the horizon."

4. above the earth: - (H1); + always (H6).

5. sun is: + on the parallel circles (H6).

[105] 1. pole: + *et nox in parte opposita* (L1).

2. faces: below (H6).

3. which faces the pole: - (L1).

4. axis: pole (H1-3,7) = diameter (H4,5). H8 reads *maḥzor* , which is apparently a mistaken transcription of *miḥwar* (axis).

5. perpendicular: + to its horizon (H6).

[106] 1. The solar...night: *et dies naturalis in eo erit medietas annis solaris* (L1).

2. Daylight: *dies vero artificialis* (L1).

3. close to: longer than (H2).

4. the northern...the sun is in: - (H4,5).

5. night...southern: - (L1).

6. because...horizon: - (H1).

7. Cancer: + or in the circle of Cancer (H6).

Chapter Five

The Orb of the Ecliptic

[107] We have previously said that each of the stars[1] has a sphere which is specifically associated with it; that every one of their [the stars'] spheres has a motion which is specifically associated with it; and that [for] each star[2], the point of its center describes an imaginary circle in its sphere[3] by means of the motion of its sphere. Thus when the orb of the sun moves from west to east, a great circle is produced in the orb of[4] the sun due to the motion of the center of the sun.

[108] We imagine this circle[1] to intersect the world. A great circle is then produced in the circle of the highest orb, whose center is the center of the world.[2] This circle is that which is called the ecliptic orb.[3] The two poles of this circle are called[4] the poles of the ecliptic orb. [109] This circle cuts the circle of the celestial equator in half at two opposite points. One of the two is called the point of the vernal equinox[1] and the other is called the point of the autumnal equinox.

[110] Again, we imagine a great circle to pass through the poles of the celestial equator, which are the[1] poles of the world[2], and through the poles[3] of the ecliptic orb. It thus intersects the ecliptic circle in two opposite points, one of which is to the north, and the other to the south. The northern [point of intersection] is called the summer solstice point, and the southern [point] is called the winter solstice point.

[111] We imagine another great circle to pass through the poles of the ecliptic circle and the two equinoctial points. These two circles divide the ecliptic circle into four equal parts. The quarter which lies between the point of the vernal[1] equinox and the point of the summer solstice is that which demarcates the season of spring. That [season] is as long as the sun, in the motion[2] of its particular orb, faces* [3] this arc.[4] It is the season which is called spring. [112] The

---

* The Arabic is *musâmitah* ; other possible translations are "points to" or "is aligned with." The Latin equivalent *respicere* is discussed by F. S. Benjamin

quarter which lies between the point of the summer solstice and the point of the autumnal equinox[1] is that which demarcates the season of summer, which lasts as long as the sun, in its motion[2]; faces[3] this arc.[4] [113] The quarter which falls between the point of the autumnal equinox and the point of the winter solstice is that which demarcates the autumnal season. That [season] is as long as the sun is facing it [that quarter] in its motion.[1] [114] The quarter which lies in that which is between the winter solstice point and the point of the vernal equinox is that which demarcates the season of winter. That [season] is as long as the sun faces it[1] [that quarter] in its motion. This entire circle demarcates the time of the solar year.[2]

[115] When the sun,[1] with its proper motion, moves from facing a point on this circle and traverses the entire circle so that it returns to face[2] that [same] point, then that motion[3] is the complete solar year. [116] The beginning of the solar year is the instant of the sun's facing the point of the vernal equinox. When it moves towards the north, it is the season of spring.[1] The closer the sun becomes to the point[2] of the summer solstice,[3] the hotter the season becomes for those living in the north, because of the nearness of the sun to their zenith, and because of what we said previously about the length of the sun's sojourn above the horizon.[4] It continues so [to be hot] until it [the sun] reaches the point of the summer solstice. The heat intensifies because of the persistence of the sun's heating of the air. [117] So it continues as it heads to the point of the autumnal equinox.[1] More heating of the air occurs in this quarter because of the length[2] [of time] of the persistence of the heating,[3] until it reaches the point of the autumnal[4] equinox, where the season returns to its equality.

[118] It then moves again, heading south. The air then cools down for those living in[1] the north. This is due to the distance of the sun from their zenith.[2] As it becomes progresses further towards the south, the coldness of the air increases,[3] until it reaches the point of the winter solstice. At that [point] the cold intensifies, because in these places the sun is at its furthest to the north, and because of the persistence of its accidental cooling of the air. [119] Then it resumes

---

and G. J. Toomer, *Campanus of Novara and Medieval Planetary Theory* (Madison and London,1971), p. 387 n. 14.

heading towards the point of the vernal[1] equinox. Again, more intensification of the cold occurs in this quarter than occurred in the previous quarter, just like that which we said about the summer quarter, [i.e.,] due to the length of the persistence of the cooling of the air.[2] When it reaches the vernal point, it is the end of its year.

[120] We also imagine two great circles drawn on the poles of the ecliptic circle. They cut the vernal quarter of the ecliptic circle into three equal parts. They[1] also cut[2] the autumnal quarter, which is opposite to this quarter,[3] into three equal parts.[4] We imagine also two other[5] circles drawn on the poles of the ecliptic circle. They cut both the winter quarter and the summer quarter opposite it into three equal parts.[6] [121] Thus the total [number] of the circles drawn on the poles of the ecliptic circle comes to six.[1] They divide the ecliptic circle into twelve equal[2] sections. They divide the highest orb [also] into twelve equal sections.[3] Two semicircles of these six circles form the boundary of each of the sections. These twelve sections[4] of the highest orb[5] are called signs.[6] The twelve sections of the ecliptic circle are also called signs.[7] [122] Each one of these arcs[1]--that is, the sections of the ecliptic circle[2]--is divided into thirty[3] equal sections by an imaginary division, each part of which is called a degree. Thus the circle in its entirety is 360 degrees.[4]

[123] We imagine the six circles which intersect the world to cut all the orbs[1] of the stars into twelve equal[2] sections similar to the sections of the highest orb, which are also called signs. [124] The ecliptic circle also divides the orbs[1] of the stars, producing[2] in each one of them a circle parallel to the great circle. Each one of these [parallel circles] is called the parecliptic orb of the ecliptic orb* . [125] These circles are also divided by the six circles into twelve equal sections, each arc of which is called[1] a sign. The orb of the fixed stars is also divided by the six circles into twelve equal[2] sections, which are also called signs.[3]

[126][1] There are many[2] stars in the orb of the fixed stars which are arranged in different patterns. In one[3] of the twelve sections there are stars from a group comprising some of the fixed stars

---

* Literally, "the orb that is similar to the ecliptic orb." This orb is so to speak the representative of the ecliptic within each set of planetary orbs. The term "parecliptic" seems to have been coined by C. Nallino. See Neugebauer, *HAMA*, vol. 3, p. 1079.

which are arranged in a pattern similar to the form of a lamb. This sign of the orb is called the sign of the lamb [Aries], and the arc of the ecliptic circle which is in the middle of this section[4] is also called the sign of the lamb. [127] This sign of the orb[1] is the first sign of the sections of the vernal quarter. It is by convention the beginning of the ecliptic circle, and the beginning of this arc is the point of the vernal equinox. Consequently, the point of the vernal[2] equinox is by convention the beginning of the circle and it is the head of Aries.

[128] In the section[1] which is adjacent to this section of the orb[2] are stars which are arranged in a form which is[3] similar to the form of an ox. This sign is called the sign of the ox [Taurus]. The arc of the circle which is in it is also called the sign of the ox.[4] [129] And so it is with all the signs: in each one[1] of them there is a form by whose name the sign is called. The arc of the ecliptic which is in that sign is called by the name of that sign.

[130] The third sign is the twins [Gemini]. The fourth is the crab [Cancer]; it is the first of the summer signs, and its beginning is the point of the summer solstice.[1] The fifth is the lion [Leo] and the sixth is the ear of grain [Virgo]. The seventh is the balance [Libra], which is the first of the autumnal signs; its beginning is the point of the autumnal equinox. The eighth is the scorpion [Scorpio] and the ninth is the archer [Sagittarius]. The tenth is the goat [Capricorn]; it is the first of the winter signs, and its beginning is the point of the winter solstice. The eleventh is the water-bearer [Aquarius]. The twelfth is the fish [Pisces]; it is the last of the signs, and its end is the point of the vernal equinox.

[131] When the sun moves with its proper motion, it always faces the ecliptic circle. If we imagine a straight line drawn from the center of the world, reaching to the center of the sun and extended rectilinearly until it reaches the highest[1] orb, it will then end up at a point of the ecliptic circle.[2] This point, then, is the point of which we say that the sun, at that instant, is residing in the sign and [in] the degree in which this point is [located].[3]

[132] As for the rest[1] of the stars, their location in the rest of the ecliptic[2] is as I shall describe.[3] Let us imagine a straight line extended from the center of the world[4] to the center of the star. We extend this line rectilinearly until it reaches the surface[5] of the highest[6] orb. It will then fall in one of the twelve signs of the

ecliptic, into which the surface of the whole[7] orb is divided. We then say that the star is at that instant in that sign.

[133] We then imagine a great circle drawn from the pole of the ecliptic circle and passing through the point which that line has reached [on the surface of the highest orb]. It is extended, passing to the other pole. It thus intersects the arc of the ecliptic circle which is within that sign. We then say that the star is in that degree which that circle has intersected in that sign. This is the position of the star in longitude.[1] [134] Similarly, if the end of the line reaches the circumference of one of the six circles, then that is the position[1] of the star.[2] The point[3] at which that circle intersects the ecliptic circle[4] is its position on the ecliptic circle; it is one[5] of the heads* of the sign.

[135] The latitude of a star is the arc of the circle passing through the poles[1] of the ecliptic circle which lies between the first point, which is the end of[2] the line passing through the center[3] of the star, and the ecliptic circle. In this manner[4] the position of all the planets and the fixed stars is found in longitude and latitude.

[136] We imagine also circles parallel to the celestial equator which intersect the ecliptic circle in the points which are the heads of the signs. These circles are called the tropics of the ecliptic,[1] because the heads of the signs move around on these circles.[2] The first of them to the north is[3] the Circle of Cancer. It touches the ecliptic circle in one point, which is the head of Cancer and the point of the summer solstice. [137] The second circle is the one which passes through the head of Gemini and[1] the head of Leo. The third[2] passes through the head of Taurus and the head of Virgo. The fourth is the circle of the celestial equator and passes through the head of Aries and the head of Libra. The fifth passes through the head of Pisces and the head of Scorpio. The sixth passes through the head of Aquarius and the head of Sagittarius. The seventh is the Circle of Capricorn and touches the ecliptic circle in one point, which is the head of Capricorn[3] and the point of the winter solstice.

[138] These tropics[1] intersect the horizons.[2] Between each two circles of the tropics of the ecliptic there is an arc of the horizon. That arc is called the rising amplitude for[3] that sign which lies

---

* The head of Aries is 0° Aries, and so forth.

between the two circles. For the orb, when is moves in its entirety,[4] sets in motion all that is contained within it. It thus sets in motions the ecliptic circle.[5] Every point on it[6] rises from a[7] point of the horizon near the east and sets from a point on it[8] near the west.[9] That point moves on a circle parallel to the circle of the celestial equator.[10] [139] The point of the head of Aries moves on the circle of the celestial equator, and the point of the head of Taurus moves on the circle of Taurus.[1] Every point of the sign of Aries and the sign of Virgo moves on a circle[2] [which] lies between these two circles, rising from the point at which this circle intersects[3] the circumference of the horizon.[4] Every point[5] of the sign of Aries and the sign of Virgo[6] rises from[7] one of the points which are on the arc of the horizon which lies between the circleof Aries and the circle of Taurus. For this reason this arc[8] is called the rising amplitude of Aries. Similarly, each one of the remaining[9] arcs which lie between the parallel circles is called the rising amplitude of that sign, and also of that sign which is with the two parallel circles.[10]

[107] 1. of the stars: of the seven planets (H6).

2. star: point (H8).

3. orb of: - (H6).

[108] 1. this circle: + the plane of (H6).

2. world: + this situation is based on [the fact that] the center of the sun always moves in the plane of the center of the world (?), which is the view of Ptolemy (H6).

3. orb: circle (all H).

4. are called: are opposite (H6).

[109] 1. equinox: - (H4, 5, 7).

[110] 1. poles . . . which are the: - (H6).

2. which are the poles of the world: - (L1).

3. the poles: the two points (all H except H6, 8).

[111] 1. vernal: - (H1,7).

2. the motion of: - (H6).

3. faces: - (H1).

4. arc: + in its motion (H7).

5. It is . . . spring: - (L1, H7).

[112] 1. equinox: *ha-hippukh* (solstice!) (H4,5).

2. motion: + special (H6) ; of its orb (H8).

3. faces: - (H1, 3, 6, 7).

4. the quarter . . . are: - (L1, H7).

[113] 1. in its motion: - (H3).

[114] 1. faces it: + this arc (H2, 4, 5, 7).

2. the quarter . . . year: - (H3).

[115] 1. when the sun: - (H3).

2. face: - (H3-5, 7).

3. motion: - (H6) = *tempus* (L1).

[116] 1. spring: *yemis* (L1).

2. point: + to the two [points] (H2-5, 7); oddly, only H5 omits "summer" in this line.

3. solstice: + *we-hu nigzar (?) me-ha-qayiṣ* (it is defined (derived?) from the summer) (H2).

4. horizon: earth (H6).

5. sun's: - (H6,8). The wording varies in the different Hebrew manuscripts.

[117] 1. So it . . . equinox: *deinde vero reditur accelerat equalitatis ad punctum autumpnalis* (L1).

2. length: - (L1).

3. heating: heat of the sun (all H save H6, 8).

4. autumnal: - (L1; H2-8).

[118] 1. those living in: - (L1).

2. from their zenith: + and due to the shortness of its sojourn above the earth (HH6, 8); their zenith: *ipsius* (L1).

3. the coldness . . . increases: The Hebrew manuscripts use different words. Significantly, H1 repeats this phrase a second time in language not found in the other Hebrew manuscripts: *we-yitpa ᶜel ha-awir me-ha-qor bi-miqreh* ; of the air: - (H6); increases: + *et augmentatur magis frigus semper propter maiorem elongacionem eiusdem* (L1).

[119] 1. vernal: - (H3-5, 7).

2. just like . . . air: *propter frigus solitum ut predictum est, et propter remocionem ipsius* (L1).

[120] 1. They: we imagine also two great circles drawn on the poles of the ecliptic [which cut . . .] (H2, 4, 5, 7).

2. into three. . . cut: *et* (L1).

3. which is opposite to this quarter: - (L1) ; of the ecliptic circle (H2, 4, 5, 7).

4. They also. . . parts: - (H3).

5. other: great (H2, 4, 5)

6. We also. . . parts: - (H6, 8); into three equal parts: - (L1).

[121] 1. six : + parts (H2)

2. equal: - (L1)

3. They divide...sections: - (L1)

4. twelve: - (H3).

5. of the highest orb: - (L1).

6. These twelve . . . signs: - (H2-5, 7).

7. The twelve. . . signs: - (L1).

[122] 1. arcs: *partium* (L1) ; the same variant is found in H2, 6-8.

2. the sections of the ecliptic circle: - (L1).

3. thirty: - (H7).

4. degrees: - (H3-5, 7).

[123] 1. all the orbs: the whole orb (H3-5).

2. equal: - (H7).

[124] 1. orbs: orb (H2, 4).

2. producing: *describit* (L1).

[125] 1. each arc of which is called: - (H6) = *que similiter nominatur* (L1).

2. equal: - (H3, 7).

3. The orb . . . signs: - (H6, 8; L1).

[126] 1. L1 is missing passages 126-132.

2. many: - (H6, 8).

3. in one: in each one (!) (all H - H6, 8).

4. section: arc (all H - H6).

5. and the arc. . . lamb: - (H3).

[127] 1. of the orb: - (H6, 8) ; of this orb (H2)

2. vernal: - (H2, 3).

[128] 1. section: Arabic Y here reads "point."

2. of the orb: of the sign (!) (H3 - 5,7) ; of the signs of the stars (H2).

3. arranged. . . which is: - (all H - H6, 7)

4. The arc. . . the ox: - (H1, 3, 7); the sign of the orb: - (H4, 5).

[129] 1. And so. . . each one: - (H4, 5).

[130] 1. and its beginning...solstice: - (H6).

[131] 1. highest: - (H6,8).

2. circle: *mi-galgal nusaḥ akher me- ᶜiggulat* (from the orb; another reading, from the circle) (H6).

3. at that instant. . . [located]: - (H2).

[132] 1. rest: + orb (H1).

2. As. . . ecliptic: - (H2).

3. As for. . . describe: *et erit collacacio stellarum in signis quod narrabo* (L1).

4. center of the world: *circulo mundi* (L1).

5. surface of the - (H1).

6. highest: - (L1).

7. whole: - (L1).

[133] 1. that sign. . . longtidude: (H2); in longitude: -(Arabic Y1, L1).

2. similarly . . . star: - (L1).

[134] 1. then that is the position: and we say that the star [is in . . .] (H3).

2. Similarly . . . star: - (L1).

3. the point: in that degree (H3 - 5, 7) = *we-ha-middah* (and the measure!) (H1).

4. Similary . . . circle: - (H2)

5. it is one: (It is its distance from one) (H2 - 5, 7): H1 also reads *merḥaqo* (its distance) but the rest of the sentence is corrupt.

[135] 1. through the poles: (in the place) (H2).

2. end of: All H except H6, 8 add here by some sort of scribal transposition, most of passage 138 (from "that sign which lies between the two circles" to "that point moves"); the Hebrew manuscripts then pick up the text as it stands in our passage.

3. center: *circulum* (L1).

4. In this manner: this is what is called (H8).

[136] 1. These circles . . . ecliptic: - (H6, 8).

2. The tropics . . . circles: - (H3).

3. is: *ex parte* (L1).

4. the head of: - (H7).

[137] 1. and: *'o* (or!) (H1).

2. third: + (circle) (H1-3).

3. and touches. . . Capricorn: - (H6, 8).

[138] 1. tropics: circles (all H - H6, 8) = horizons (!) (H6, 8). See following note.

2. horizons: orbs(H2, 4, 5, 7).

3. That arc. . . for: - (H4, 5).

4. in its entirety: in its orb (H7).

5. It thus . . . circle: - (L1).

6. on it: *circuli signorum* (L1).

7. from a: opposite (H6, 8).

8. on it: on the point of the horizon (H2).

9. west: *meridiei* (L1).

10. the point. . . equator: (H2).

[139] 1. moves . . . Taurus: - (H7).

2. on a circle: - (Arabic Y).

3. intersects: *ta* ^{c}*aleh* (rises) (all H except 6).

4. rising . . . horizon: - (L1).

5. point: - (H3, 5, 7) ; arc (H2).

6. Every point . . . Virgo: *punctus* (L1).

7. from: ^{c}*im* (with) (H6).

8. this [arc]: the head of (H6).

9. remaining: - (H4,5).

10. of that sign. . . circles: *illis duobis interiacentibus duos circulos interiacentes equidistantes* (L1).

Chapter Six

The Declination

[140] The declination is an arc of a great circle which passes through the poles of the world, between the ecliptic circle and the circle of the celestial equator. [141] Every point of the ecliptic circle[1] has a declination from the circle of the celestial equator, except for[2] the two points of intersection, which are the heads of Aries and Libra. [142][1] This is because all the circles which are drawn on the poles of the world to every point on the ecliptic circle intersect it [the ecliptic] and reach to the circle of the celestial equator. An arc of the circle which passes through the poles is thus produced between each point of the ecliptic circle and the celestial equator.[2] This arc, then, is the declination of that point.

[143] These declinations are different. The closer [one] of them is to the two points of intersection of the ecliptic circle and the circle of the celestial equator[1], i.e., to the two equinoctial points, the smaller is [the declination]. The further [one] of them is from these two points, the greater is [the declination]. [144] The greatest declination is the declination of the point[1] which is in the middle of the semicircle of the ecliptic,[2] namely the point of the solstice.[3] The greatest declination is an arc of the circle which passes through the poles of[4] the world and [through] the two solstitial points,[5] between the solstitial points and the circle of the celestial equator. This arc is approximately twenty-four[6] of the parts of which 360 constitute the whole circle.

[145] The arcs of those circles which pass through the points[1] that demarcate the signs are called the declinations of[2] those signs. As for the [circle] which passes through the head of Taurus and the head of Scorpio, the arc on it which lies between the head of Taurus and the celestial equator is called the declination of Aries, and its opposite [arc] is called the declination of Libra. [146][1] The arc of the circle which passes through the head of Gemini and the head of Sagittarius is called the declination of Scorpio[2] and Taurus. [147][1] That[2] which passes through the head of Leo and the head of Aquarius is called the declination of Cancer[3] and Capricorn.[4] [148]

That which passes through the head of Virgo and the head of Pisces is called the declination of Leo[1] and Aquarius.[2] [149] That which passes through the points which lie between these points is called the declination of that degree, that is, that [degree] which that point demarcates. All of these declinations are equal to the declinations of the points opposite them.[1]

[150] Every point of the ecliptic circle moves, as we said before, on circles parallel to the celestial equator. The distance of that circle from the celestial equator is the declination of that point. When the sun is facing a point of the ecliptic circle, it is moving in the plane of the circle parallel to the celestial equator which that point describes. Thus its declination from the celestial equator is always the declination of the point of the ecliptic circle[1] which it is facing.[2]

[151] Because, as we have already shown, the latitudes of the inhabited places are arcs lying between the zenith[1] and the celestial equator of the circles which pass through the poles of the world and the zenith points,[2] it necessarily follows from that that the circles of declination correspond in the motion of the whole to the circles of latitude.* Thus the declination of each point of the ecliptic circle is equal to one of the latitudes. Thus that point, in the motion of the whole, corresponds to[3] the zenith point for that latitude in each revolution. [152] For this reason,[1] the sun, when it is facing that point of the ecliptic circle [whose declination is equal to a given terrestrial latitude], passes on that day through the zenith of all who are in that latitude. It follows from that that the sun passes through the zenith of those living in places[2] whose latitude is not greater than the maximum declination.[3] [153] For those whose habitation has a latitude equal to the maximum declination, which is approximately twenty-four[1] parts,[2] the sun passes through the zenith[3] when it is in the solstitial point.[4] At that time their gnomons do not have a shadow at noon of that day. As for the rest of the year, the shadow of the gnomon is to the north for that habitation.[5]

[154] For those habitations whose latitude is less than the maximum declination, the sun passes through the zenith twice every

* The correspondence indicated here is that between terrestrial latitude and declination. Ibn al-Haytham will explain verbally that only a star having a declination equal to a given locality's latitude (even if only temporarily, as in the case of the sun) will pass through the zenith point of that locality.

year, because the circles parallel to the celestial equator which intersect the ecliptic circle intersect it in two points. The declination of those two points[1] is equal, because it is the distance of that circle from the celestial equator. When the sun reaches the first of[2] these two points, it is moving in[3] the plane of that circle, passing through the zeniths of those whose latitude is equal to the declination of that degree [of the ecliptic]. Then their gnomons do not have a shadow on noon of that day. [155] When the sun moves from that point to the north, it goes away from their zeniths, and the shadows of their gnomons[1] are to the south. It continues like this until it [the sun] reaches the second[2] point, and then again[3] it passes through their zenith. When it moves from that point towards the south, the shadows of their gnomons are to the north.

[156] As for those whose latitudes are greater than the maximum declination, the sun does not pass through their zeniths, but rather it is always inclined to their south. The shadows of their gnomons[1] are always to the north. This that we have related is the situation with regard to the declinations of the sun.

[157] The declination of the planets and fixed stars is an arc of a great circle between the point facing the center of the star and the celestial equator, [the great circle being] either from among the circles which are drawn on the poles of the world[1]--the arc is then called the declination of the star--or[2] from among the circles which are drawn on the poles of the ecliptic* circle,[3] and then the arc is sometimes called the declination of the star and sometimes it is called the latitude of the star with reference to the celestial equator.

---

* Unless there is a textual problem which has entered into all of our manuscripts, we have here a new type of coordinate. Now in addition to the standard ecliptic and equatorial systems, medieval astronomers knew of and used a system of mixed coordinates, in which the "polar latitude" is the distance of the star from the ecliptic measured along the circle passing through the star and the pole of the equator; see Neugebauer *HAMA* , vol. 3, pp. 1081-1082, and Y. T. Langermann, P. Kunitzsch, and K. A. F. Fischer, "The Hebrew Astronomical Codex Sassoon 823," *Jewish Quarterly Review* 78(1988) 253-292, pp. 272-275. However, Ibn al-Haytham is here referring to what may called the converse of this system, i.e. using the circle through the star and and the pole of the ecliptic, and on this circle measuring the distance of the star from the celestial equator.

[158] The declinations of the planets vary, as the declination of the sun also varies, because each one of them has a circle specific to it intersecting the circle of the celestial equator; their motion is according to its reckoning. It [the planet] has a declination to the north and to the south [just] as the sun has a declination in these two directions. Also, the more distant it is from the two points of intersection [of its circle and the celestial equator], the greater its declination. [159] The declination of the fixed stars does not vary discernibly because they move slowly, but over a long period of time differences in their declinations become evident.

[141] 1. circle: + orb (L1), yielding the reading " the circle of the ecliptic orb".

2. Every point. . . except for: to (H6).

[142] 1. This passage is missing in L1.

2. between. . . equator: - (H1).

[143] 1. of the ecliptic. . . equator: - (L1).

2. semicircle of the ecliptic: these points (H2- 5, 7).

3. The greatest . . . solstice: - (H6); solstice: + summer (H1); (note that this variant occurs also in Arabic K); *tropici estivi et puncti tropici yemalis* (L1).

4. the poles of: - (H6).

5. The greates. . . points: - (H3).

6. twenty-four: *quasi 23 graduum et 51 minutorum* (L1). This is in fact Millás' correction; he notes that instead of 23, "El manuscrite dice: 13."

[145] 1. which pass through the points: - (H7).

2. the declinations of: - (H3).

[146] 1. This passage is missing from H2- 5, 7.

2. Scorpio: *ṭaleh* (H1, 6) = *arietis* (L1). Note that this follows the reading in Arabic Y.

[147] 1. This passage is missing from H6.

2. that: *ha-qeshet* (the arc) (H1, 2).

3. Cancer: Leo (all translations).

4. Capricorn: L1 reads "Virgo," as do H1 and H8; this follows the reading of Arabic Y; H2- 5, 7 read Aquarius following Arabic K.

[148] 1. Leo: All Latin and Hebrew manuscripts follow the Arabic text, reading "Virgo."

2. and Aquarius: The Latin manuscripts and H3 -5, 7, omit this, thus following Arabic Y; H1, 2, 6 and 8 read " and Pisces" as in Arabic K.

[149] 1. All . . . them: - (H2- 5, 7); opposite them: - (H1).

[150] 1. it is moving . . . circle: - (H2- 5, 7).

2. which it is facing: which passes through the zenith (all H except H6, 8, with minor variants in H2, 3).

[151] 1. between the zenith: - (H6, 8).

2. zenith points: *nokhaḥ ha-shemesh* (facing the sun; *ha-ro'sh* , which would give the phrase the same reading as out text, is crossed out, and *ha-shemesh* is substituted) (H2).

3. Thus the declination. . . corresponds to: *ᶜal* (on) H2 - 5, 7).

[152] 1. For this reason: - (L1)

2. those living in: - (L1).

3. those living in . . . declination: - (H1).

[153] 1. twenty-four: *23 graduum et LI minutorum* (L1; again, Millás notes "et manuscrito dice: 13 graduum").

2. parts: *ma ᶜaleh* (degree) (H2 - 5, 7) = (Arabic K).

3. for those . . . zenith: - (H1).

4. solstitial point: *punctum capricorni* (L1); Arabic K adds here "summer."

5. At that time. . . habitation: *et tunc non faciunt umbram, et cum sol moveatur ab aliquo puncto ad alium, parte meridiei, et nunquam habent umbram versus septentrionem* (L1).

[154] 1. The declination of those two points: - (H2,3, 6).

2. the first of: - (L1).

3. in: *ᶜal nokhaḥut* (all H - H6, 8) = *verticando* (L1).

[155] 1. are: *redeunt* (L1); the Arabic here is *ṣârat* .

2. second: *alterum* (L1).

3. again: - (L1).

[156] 1. of their gnomons: - (L1).

[157] 1. either. . . world: - (H3).

2. either. . . or: - (H6, 8).

3. circle: or: - (all H).

[158] 1. intersecting: *nokhaḥit* (facing) (all H - H6, 8).

2. It. . . directions: the star is inclined in any direction (H6, 8).

Chapter Seven

The Altitude

[160] The altitude is an arc of the zenith[1] circle between the point which has risen[2] and the horizon. [161] When each star rises above the horizon, the measure of its altitude is the arc of the great circle that is drawn from the zenith, passing through the point facing the center of the star and reaching to the horizon.[1] I mean by "the point facing" the point[2] which is the extremity of the line drawn from the center of the world[3] to the center of the star and which, [extended] rectilinearly, reaches the surface of the highest orb. [162] As the altitude of the star increases, the arc becomes greater. It continues like this until it [the star] reaches the meridian circle. Then the arc of the meridian circle--which is also one[1] of the zenith circles--between the point facing the center of the star and the horizon is the limit[2] of its altitude. [163] Then is goes down to the west, and its altitude decreases because this arc becomes smaller. So it continues: as it approaches the west, its altitude decreases, until it reaches the western point. Such is the case[1] of[2] the sun and all the stars[3] which rise[4] and set.

[164] The stars which are always visible move on visible* circles which are always visible,[1] but their altitudes also vary. For they move upwards,[2] and their altitude increases, until they reach the meridian circle[3]; there they are at their greatest altitude. Then they move downwards and again arrive at the meridian circle, because the meridian circle intersects the circle on which the star moves in two visible points. When the star in its descent reaches the meridian circle,[4] there and then is its least altitude. This is what is called the altitude.

[165] The true altitude is the perpendicular dropped from the center of the star to the plane of the horizon. This perpendicular is the sine of the arc parallel and equivalent to the arc of altitude which

---

* I.e. above the horizon.

passes through the center of the star and reaches to the horizon. [166] The sine of every arc[1] is half the chord of double that arc.[2] This is the definition of[3] the "erect" sine. [167] As for the "versed" sine, it is the "arrow" of[1] double[2] the arc.* Sines are used in many situations, because from them the arcs are calculated. The greatest altitudes of the stars are the sines of their meridians.** [2]

---

* In modern formulation, vers $\phi = 1 - \cos \phi$. For information on some of the medieval applications of this function, and the differing terminologies, see M.-T. Debarnot, *Al-Bîrûnî: Kitâb Maqâlîd ᶜilm al-Hay'ah. La Trigonometrie sphérique chez les Arabes de l'Est à la fin du Xe siècle* (Damascus,1985), pp. 228-29; E. S. Kennedy, *A Commentary upon Bîrûnî's Taḥdîd al-Amâkin* (Beirut,1973), p. 23; B. R. Goldstein, *The Astronomical Tables of Levi ben Gerson* (Hamden, Ct., 1974),p. 54; see also our lexicon s.v. *jayb* . "Arrow" is the literal translation of the Arabic *sahm* , Hebrew *ḥeṣ* , and Latin *sagitta* .

** The meridian of the star passes through the star and the celestial pole, just as the meridian of an observer passes through his local zenith and the celestial pole. This terminology is still in use, as we see from the revised edition of a standard textbook, W. M. Smart's *Textbook of Spherical Astronomy* (sixth edition, revised by R. M. Green, Cambridge, U.K.,1977, p. 30): "When the star is in transit, its meridian coincides with the observer's meridian."

[160] 1. zenith: *centralis* (L1). Note that here Arabic K reads "solar."

2. has risen: + that is, the point of the head (!)(H2).

[160] 1. horizon: + *et ille arcus terminatur ad punctum verticantem centro stelle et orizontem* (L1).

2. the point: - (H2 - 5, 7). The relative "that" is understood in the Hebrew.

3. world: ^{c}*iggulah* (circle) (H6).

[162] 1. one: + *davar* (thing)(!) (H2).

2. limit: *maxima* (L1).

[163] 1. case: *ha-zorḥim* (the bright bodies) (H1). This ms. obviously meant this term to be an addition to, rather than a replacement for. the word "case" in our text.

2. case of: + *govah* (the altitude of) (H6).

3. stars: + *ha-po* ^{c}*alim* (active) (H6). In H8 the same word replaces "which rise."

4. which rise: H2, 3, and 7 read ^{c}*iggulim* (circles), a change of one letter from ^{c}*olim* (which rise).

[164] 1. move. . . visible: - (L1). [always] visible: *n* ^{e}*gohim* (?) (H1). The reading of this word is uncertain. It has the connotation of "bright".

2. For. . . upwards: - (H6, 8).

3. there. . . circle: - (H3, 7); again. . . circle - (H4, 5).

4. because the meridian. . . circle: - (L1).

[166] 1. [every] arc: '*ofeq* (horizon) (H6, 8).

2. that arc: + *ad arcum cuius est sinus* (L1).

3. the definition of: - (H6, 8)

[167] 1. the axis of: - (H2).

2. of double: - (L1).

3. the sines of their meridians: when they are not in their meridian orb (H2 - 5, 7) = is the sine of double the full arc of the meridian (H6, 8).

Chapter Eight

The Ascendant and the Ascensions[1]

[168] The ascendant is the part* of the ecliptic circle which is on the circumference of the horizon[1] to the east. The descendant is the part opposite it to the west. For the horizon cuts the ecliptic circle in half, and there are always on the circumference[2] of the horizon two diametrically opposite[3] points of the ecliptic circle. [169] For this reason, whenever a part of the ecliptic circle rises,[1] the part opposite it sets. So it is for each of the great circles which are in the orb.

[170] This is not the case for the circles on which the fixed stars move nor for the rest of the parallel circles. That is to say the part of them which is in the east is not opposite the part[1] which is in the west for all the horizons. For the inclined horizons do not cut all these circles in half, but rather in different sections, except for those of them which are great circles.[2]

[171] The ascension is the arc of the celestial equator[1] which rises along with a given arc of the ecliptic circle.[2] [172] We will give an example of this, which is[1] [the following]. We assume the point facing the center of the sun to be on the circumference of the horizon.[2] In this case, a certain point[3] of the celestial equator is on the circumference of the horizon.[4] We then imagine the sun[5] to be moving upwards. Then the [corresponding] point of the celestial equator is [also] rising. Then, when the sun reaches a certain given hour** [6] of the ecliptic circle on which it moves, an arc of the ecliptic circle has already risen in that amount[7] of time, namely that[8] [arc which extends] from the point facing the center of the sun to the eastern horizon.[9] At that time the point of the celestial equator[10] has already reached a certain part of the celestial equator. The arc of the celestial equator[11] which lies between that part and the horizon is the ascension of that arc of the ecliptic circle.[12]

---

* Here the Arabic *juz'* could just as readily be translated "point."

** An "hour" is a unit of 15°. The reference here is to the sun's day circle.

[173] This arc[1] of the celestial equator[2] is equivalent to the arc of the circle parallel to the celestial equator[3] facing which[4] the center of the sun moves. That arc is also called the ascension of that arc of the ecliptic circle.[5] I mean by "equivalent" that the number of parts of the two is equal.[6]

[174] For each part of the ecliptic circle there is[1] a part of the celestial equator which is the ascension of that part.[2] The ascensions of the parts of the ecliptic circle are different, with this exception[3]: each two parts of it [the ecliptic] which are equal--that is, whose distance[4] from one of the equinoctial points is the same-- will have equal ascensions. For example, the first ten degrees of Aries and the last ten degrees of Pisces have equal ascensions. [175] So it is for the whole of Aries and the whole of Pisces. So also the ascensions of Virgo and Libra are equal,[1] and so it is for all the parts of the ecliptic circle[2]: the ascensions of each quarter are equal to the ascensions of the adjacent quarter. The ascension of each part of the quarter is equal[3] to the ascension of the corresponding[4] part of the adjacent quarter.[5]

[176] This is the situation of the ascensions with regard to each one of the horizons. However, the ascensions of each part of the ecliptic circle in a given horizon are different from the ascensions of that part in a different horizon.[1] The difference in the ascensions is according to the difference in latitude. [177] As an example of this, the ascension of Aries for a given horizon is a given [number of] parts. Its ascension in the horizon whose latitude is greater than that latitude is a certain [number of] parts[1] less than those parts [of the first ascension]. It is likewise for Pisces[2]; contrariwise* for Virgo and Libra.[3] The ascension of the semi-circle[4] of the ecliptic which lies between the two equinoctial points is always half the circle of the celestial equator.

[178] As for the horizon[1] whose latitude is 66 parts, which is the one whose longest daylight is 24 hours, and to whose horizon[2] the Tropic of Cancer is tangential--in this horizon the ecliptic circle

---

* The rising times of Virgo and Libra increase as the northerly latitude increases.

coincides with the horizon circle for an entire day and night*. [179] When it begins to move after coincidence, six signs rise simultaneously.[1] Hence there is no ascension for this half [circle] in that horizon**.[2] Then the orb begins to move, and the situation of the ascension in that horizon reverts to its situation in the rest of the horizons.

[180] No part of the ecliptic circle rises or sets in the place which is under the pole, but rather half of it is always visible and half of it is always hidden.[1] [181] As for the places[1] which are between the place which is under the pole and the horizon whose latitude is 66, part of the ecliptic[2] is always[3] visible there,[4] and it does not have an ascension in those places. The corresponding[5] part of the the ecliptic is always hidden, and it [also] does not have an ascension in those places.[6] [182] That[1] part of the ecliptic which is between the [eternally] visible part and the [eternally] hidden parts rises and sets. Its ascension it like the ascension of other [parts] in the rest of the horizons.[2] Every arc of it rises along with an arc of the celestial equator.[3] That [part] of the ecliptic which rises and sets in these horizons is two non-adjacent arcs. The visible part separates the two [rising and setting parts] from one of the extremities, and from the other extremity the hidden part[4] [separates the two]. One of these two [rising and setting] arcs is near[5] the point of the vernal equinox, and this point cuts it in half. The other is near the point of the autumnal equinox, and this point cuts it in half.[6] [183] The rising of the arc[1] which is near the vernal equinox on these horizons[2] is inverted and its setting regular.[3] The rising of the arc which is near the autumnal equinox is regular and its setting is inverted.[4] This is the whole section on the ascensions.

[184] Since we have finished our commentary on these imaginary circles in the surface of the highest orb, we will draw a diagram of this in which we will record all the circles[1] and lines which we have

---

* So read both Arabic manuscripts. The Alfonsine translation (variants from which are not listed in the apparatus) replaces "every" with *semel* , thus implying the correct statement that once each year, the ecliptic will coincide with the horizon for an entire nychthemeron.

** We must here accept the version found in most mss. of the Hebrew translation; cf. the apparatus.

mentioned, so that all that is in the surface of the orb may be represented to the reader[2], and this is the figure.

[185][1] We have already spoken of [the fact] that each star has its own altogether[2] particular[3] orb which is part of the whole encompassing body. Each of the orbs[4] of the planets[5] is divided into parts. We shall discuss the case[6] of each planet and its orbs.[7] We shall begin with the noblest of them, which is the orb of the sun. It will be followed by those which are adjacent, one by one.[8]

Chapter Title

1. *De ascendente et descensionibus* (L1).

[168] 1. of the horizon: the whole (H6).

2. always on the circumference: - (H6, 8).

3. diametrically opposite: - (L1).

[169] 1. For . . . rises: - (H4, 5).

[170] 1. the part: - (H2, 4, 5, 7).

2. but rather. . . circles: - (L1).

[171] 1. arc. . . equator: *circuli arcus* (L1).

2. circle: + *in* (?) *gradus* (L1).

[172] 1. We. . . is: - (L1).

2. Circumference of the horizon: *ha-maqif* (the encompassing) (H6, 8).

3. of: *she-ben* (which is between) (H2, 3, 5); *ben* (between) (H4, 7).

4. on the circumference. . . horizon: also on the horizon (H6, 8).

5. the sun: - (H2, 5, 7).

6. a certain assumed hour: - (H4, 5).

7. amount: *sha ᶜah* (hour) (H1).

8. that: + arc (H6, 8).

9. to the eastern horizon: - (L1). Millás notes: "el resto de la linea está en blanco."

10. At that. . . equator: - (H3, 7).

11. has already. . . equator: - (H2, 4, 5); has already risen also, therefore (H6, 8).

12. which lies . . . circle: - (H6).

[173] 1. arc: - (H3).

2. this . . . equator: - (H6).

3. of the circle. . . equator: H2- 5, 7 transpose "circle" so that the sentence reads: the arc parallel to the circle of the celestial equator.

4. facing which: *ᶜaleha* (on which) (H6, 8).

5. circle: + *galgal* (orb) (H1). The phrase thus reads "the circle of the orb of the ecliptic." Note that Arabic K has "orb" written above "circle" in the manuscript.

6. I mean . . . equal: - (H2, - 5, 7); that the number. . . equal: *quod medius partium eius equator medio parcium illius cuius est simile* (L1).

[174] 1. For each . . . is: - (H2 - 5, 7).

2. For . . . of that part: - (L1).

3. with this exception: that is (H7).

4. distance: *z ᵉman ḥuqehem* (their appropriate time[s]) (H4, 5). This seems to be a misreading of *merḥaqehem* (their distance).

[175] 1. So. . . equal: - (H1, 2, 7; L1). Note that Arabic K is also missing this section.

2. So it is for. . . circle: - (L1).

3. Ascension of the adjacent . . . equal to: - (H2 - 5, 7).

4. corresponding: *ha-n ᵉkhoḥi* (the opposite) (all H except 6).

5. The ascension of the. . . quarter: - (H6, 8). Thus this sentence is found in H1 alone among the Hebrew translations; *et cuiuslibet partis parcium quare ascensiones equantur ascensionibus partis* *anadir* *parti continuate* . (L1). Millás has the following note on the word anadir: "Esta nota marginal explica la palabra anadir: *Sunt partes sese respicientes ut prima ultimum et e converso, post prima penultimam, tercia post priman antepenultiman, et e converso, etcetera* ."

[176] 1. This . . . horizon: - (H7); However. . . horizon: - (H1).

[177] 1. certain [number of] parts: - (H6, 8).

2. Pisces: + the ascensions of the sign of (H6).

3. Virgo and Libra: *b ᵉtulah u-mo'znayim bi-re'shit shel mo'znayim* (Virgo and Libra in the beginning (first degrees] of Libra) (H1) = *b ᵉtulah sofo mimennu bi-reshit shel mo'znayim* (for Virgo, whose end is in the beginning of Libra) (H3, 7). The texts in H2, 4, and 5 read like H3 and 7, but substitute *sofo ha-aḥaron* (its very end) for *sofo* (its end). Only H8 reads like the Arabic.

4. semi-circle: the part of the circle (H6, 8).

[178] 1. horizon: *oriente* (L1). An obvious error for *orizonte* .

2. horizon: - (H1).

[179] 1. simultaneously: in no time, i.e., (H6).

2. Hence. . . horizon: - (L1); horizon: *ha-* ^{c}*et* (instant) (all H except H6, 8).

[180] 1. always. . . hidden: - (H6, 8).

[181] 1. the places: in the remaining (H6, 8).

2. of the ecliptic: of the circle, i.e., the ecliptic circle (H2 - 5, 7).

3. always: - (H6, 8).

4. there: + and the opposite [part] of the ecliptic circle is always hidden and has no ascension in those places (H8).

5. corresponding: nadir (L1). An apparent transliteration without translation of the Arabic *naẓîr* .

6. The corresponding. . . places: - (all H).

[182] 1. That: *Et illiud quod interiacet primas duas partes scilicet* (L1).

2. The horizons: H7 has this word in the singular.

3. Its ascension. . . equator: - (L1)

4. hidden part: + *ha-ḥeleq ha-nifradi* (the separate part) (?) (H2).

5. One. . . is near: *w e-eḥad mehem ben 'elu ha-q eshatot* (one of them is between these arcs) (all H except 6).

6. One. . . half: (L1); The other. . . half: - (H2 - 5, 7: L2).

[183] 1. the arc: *w e-asher* (that) (H6, 8).

2. one these horizons: *b e-elu ha-hippukhim* (?) (H8). *Hippukh* is the word used for "solstice".

3. regular: + *et e converso* (L1).

4. The rising. . . inverted: - (H6, 8).

[184] 1. in the surface. . . circles: - (H7).

2. so that. . . reader: - (L1).

[185] 1. Arabic K is missing this passage; H1, 6, and 8 add here a chapter title: "the chapter of the orbs of the stars." This seems to be a demarcation between the two parts of this book.

2. altogether: - (H6, 8: L1).

3. special: one (H6, 8).

4. orbs: + 7 (L1).

5. planets: *ha-kokhavim ha-ḥolqim* (the stars which divide) (!) (H7). Apparently, this text was dictated; the reader must have read *holkhim* .

6. case: *astrologiam* (L1).

7. orbs: All H except H6, 8 have this word in the singular.

8. It will. . . one: Afterwards we will talk about each one of the remaining [stars] (H6). H8 has a similar reading, adding "one by one."

Chapter Nine

The Orb of the Sun

[186] The orb of the sun is a round[1], spherical body bounded by two parallel spherical surfaces whose center is the center of the world. The higher[2] of the two surfaces is convex and[3] tangent to the concavity of[4] the orb of Mars. The lower of the two surfaces is concave and[5] tangent to the convexity of the orb of Venus. [187] A round sphere is separated from it [the orb of the sun]. It is also bounded by two parallel[1] spherical surfaces, whose center is[2] a point removed from the center of the world. The upper surface of this sphere is tangent to the upper surface of the first sphere in one point common to the two of them.[3] Its lower surface also is tangent to the lower[4] surface[5] of the first sphere[6] in one point common to the two of them.[7] This orb is called the eccentric orb. It moves with a circular motion from west to east on two fixed poles and a fixed axis[8], which are different from the poles and the axis[9] of the world and different from the poles of the ecliptic circle and its axis* .

[188] The sun is a solid[1] body whose shape[2] is that of a sphere[3], fixed in the body of that orb.[4] Its [the sun's] center is equidistant from its [the orb's] poles. Its [the sun's] exterior surface[5] is tangent to its [the orb's] two surfaces in two points. I mean that two points on the surface[6] of the body of the sun, namely the two extremities of its diameter, are common to its [the sun's] surface and the two surfaces[7] of its second orb. The diameter of the sphere[8] of the sun is thus equal[9] to the width of the body[10] of its orb. When this orb moves from west to east on its fixed poles, its sets in motion the body of the sun along with it[11], because it [the latter] is like a part[12] of it.

[189] When[1] the sun moves with its circular motion, the point of its center describes an imaginary circle whose center is the center of

---

* Different from the poles and axis of the ecliptic in that they are physically distinguishable, since they are located on or within different orbs. The axes of the sun's eccentric orb and the ecliptic orb (drawn on the greatest, outermost sphere) are collinear. See passages 189 and 190 below.

this orb. This circle is also called the eccentric orb.[2] [190] If we imagine the plane of[1] this circle to intersect the world, the result is a circle in the surface of the highest orb, which is the ecliptic circle, as we explained earlier. There also result two parallel[2] circles[3] in the two surfaces[4] of the orb of the sun whose center is the center of the world, the center of which [i.e., these two circles] is the center of the world.[5] The larger is called the parecliptic orb.

[191] Two mutually parallel circles, parallel to the eccentric orb[1], are produced in the two surfaces of the eccentric orb of the sun. The larger of the two is tangent[2] to the parecliptic orb. The smaller is[3] tangent to the circle parallel to it.[4]

[192] In each of the orbs of the stars by which we initially divided the entire orb into nine[1] mutually contiguous divisions whose center is the center of the world, two parallel circles[2] whose center is the center of the world result. The larger of them is called the parecliptic orb of that star, and the smaller of them is coincident with the larger one of the adjacent[3] orb.

[193] The two tangent points[1] of the eccentric orb are the furthest distance and closest distance of the eccentric orb.[2] That is, one of them is the point on the circumference of the eccentric orb[3] furthest from the center of[4] the world, and the other is the point of its circumference closest to the center of the world.[5] For if a straight line be drawn between the center of the world and the center of the orb of the sun[6] and extended rectilinearly in both directions, it will reach the tangent point.[7] The segment which is near the eccentric center is the longest line which [can be] drawn from the center of the world to the circumference[8] of the eccentric orb[9], and the remaining segment[10] is the shortest of the lines drawn from the center of the world to its circumference.[11] [194] This line intersects the circumference of the eccentric circle which passes through the center of the sun and which is also called the eccentric orb[1] in two opposite points. That[2] [point] which is on the longer line is also called the furthest distance[3] and that[4] which is on the shorter line is called the closest distance.[5] The point of furthest distance is called the apogee of the sun[6], and the point of closest distance is called the perigee of the sun.[7]

[195] The apogee point faces[1] a point of the ecliptic orb[2] to the north whose distance from the summer solstitial point is 24 1/2

parts[3] of the parts of which 360 make up a circle, preceding it[4], as Ptolemy noted in his book known as[5] the *Almagest* .* [6] [196] According to what he noted, this point is fixed and does not change position from facing the point of the ecliptic orb[1] whose distance from the summer solstice is this [same number of] parts. [197][1] The perigee point precedes the point of the winter solstice to the south by a like [number of] parts.[2] [198] According to what the later astronomers say, the two of them [apogee and perigee] move and change position[1] in the order of the signs** .

[199] When the sun moves with the motion of its orb in a circular motion, its center continually faces the parts of the ecliptic circle, part after part.[1] That is, the lines drawn from the center of the world to the center of the sun reaching to the ecliptic circle in successive instants will reach successive points[2] [of the ecliptic], and continue to do so until it[3] [the sun] comes back to face the first point.[4] Then the sun's year has been completed.[5]

[200] Thus the sun always[1] keeps to the circumference[2] of the eccentric orb.[3] When it is in the furthest distance of its orb, i.e., the apogee, it is as far as possible from the earth. It[4] then changes its position[5] and continues to come nearer[6] [to the earth] until it reaches[7] the perigee of its orb.[8] At that time it is as close as possible to the earth. It then continues to ascend[9] until it returns to[10] its apogee. Thus it is always.[11] Because of this situation, its distances from the earth vary.

[201] It [the sun] thus faces equal arcs of the ecliptic circle in unequal times, that is, it passes through the parts of the semicircle of the ecliptic which is near the apogee in an [amount of] time greater than the [amount of] time in which is passes through the semicircle [of the ecliptic] which is near the perigee.

[202] In this manner the sun has two motions. One of them is its proper motion[1], which is the motion of its eccentric orb[2] about its

---

* Book 3, ch. 4; G. J. Toomer, *Ptolemy's Almagest* (New York-Berlin-Heidelberg-Tokyo, 1984), p. 155.

** For Ptolemy's view see *Almagest, loc. cit.,* and Toomer, p. 153, n. 46; concerning the later astronomers, see W. Hartner and M. Schramm, "Al-Bîrûnî and the Theory of the Solar Apogee: an example of originality in Arabic science," in A. C. Crombie (ed.), *Scientific Change* (New York,1963) 206-18.

own center and on its own poles, that is, the equal motion.[3] The other is that which exists in relation to the ecliptic circle, whose center is the center of the world. It is the unequal[4] motion.

[203] Since the sun, when it is in its perigee, is as close as possible to the earth, it necessarily follows that the place[1] on the earth facing its perigee be the hottest place* . This place on the earth[2] is to the south, close to the line called the terrestrial equator, facing the point which we previously mentioned, and it[3] girds the earth. [204] For in the time when the sun is in its perigee,[1] the highest orb sets it [the sun] in motion with the swift motion. Thus it [the sun] circles around the earth in a day and a night[2] on a circle parallel to the celestial equator, it [the sun] being in its closest proximity.[3] It thus also faces a circle[4] on the earth which is the closest circle to that circle.[5]

[205] That is, if we imagine a straight line drawn from the center of the world[1] to the center of the sun when it is in its perigee[2], then it [the line] meets the surface of the earth in a point. Then, as the sun moves around the earth, we imagine that line to move with it, and thus the point on it [the line] which is on the surface of the earth also produces a circle on the surface of the earth.[3] It is not a great circle[4], because the time[5] circle on which the sun moves when it is in its perigee is not a great circle. [206] The straight line drawn to it [the sun] from the center of the world which moves with it [the sun] does not move on a flat plane, but rather on a conical** [1] surface. For

---

* Medieval geographers reasoned that the heating of the earth's atmosphere due to the sun is due to two variables: the seasonal variation, on account of the changing solar declination, of the sun's maximum altitude; and the sun's changing distance from the earth. In the northern hemisphere, summer occurs when the sun is in the furthermost part of its orbit from the earth; thus its distance at that time moderates the effect of the greater altitude. For the southern hemisphere, however, the two factors- greatest altitude and closest distance- reinforce each other, producing what was held to be an intolerable level of heat. It may be added that current theory teaches that the sun-earth distance is not an important factor; winds and currents tend to even out the distribution of heat over our planet. Nevertheless, there may be some overall net difference in the average seasonal temperatures between the two hemispheres.

** Concerning some imprecise medieval usages of the term "conical" (see lexicon, s.v. *makhrûṭ*) see F. S. Benjamin and G. J. Toomer, *Campanus of*

this reason the section of the sphere of the earth about which it moves[2] is not[3] a semi-sphere. Thus the circle which is produced from its [the line's] motion[4] on the surface of the earth is not a great circle. [207] Because that line, that is, the line which is drawn from the center of the world[1] to the perigee of the sun[2] is the shortest line which [can be] drawn to the sun, and [because] the radius of the earth does not vary, the segment of it[3] [the line] which is between the surface of the earth and the perigee[4] is the shortest line that can be found during the length of the year[5] between the center of the sun and the surface of the earth.[6] Hence this circle which we have mentioned[7] is the closest [to the earth] of the circles of the sun. Because of this[8], it is the hottest place on earth.[9]

[208] As one approaches it [that place], it is also intensely hot.[1] This circle[2] is called the "burning path."* For this reason, the southern direction is hotter than the northern direction. The equator of the earth is also intensely hot due to its proximity to this path.[3] It is said[4] that these places of the earth are uninhabited due to the intensity of the heat, and that there is no living being[5] in them.

[209] Let us draw a diagram for[1] the orb of the sun according to[2] the circle which is produced by the motion of the sun and for all the circles which it produces.[3] This is it.[4]

Novara and Medieval Planetary Theory (Madison and London,1971), p. 383 n. 23.

* It seems that this is not the same as the "burning path" known to astrology. See al-Bîrûnî, *The Book of Instruction in the Elements of the Art of Astrology (al-Tafhîm)* , trans. R. Ramsay Wright (London,1934), p. 317.

[186] 1. round: - (L1).

2. higher: - (H4, 5).

3. convex and: - (L1).

4. the concavity of: - (H7).

5. concave and: - (L1).

[187] 1. spherical: - (L1).

2. center is: + *merkaz* (a center) (H2, 3, 4, 6-8). The purpose of the repetition of this term is unclear.

3. The upper. . . of them: - (H2); in one. . . of them: - (H3 - 5, 7).

4. lower: upper (H4, 5)

5. is tangent to the lower surface: - (H7).

6. of the first sphere: - (L1).

7. Its lower. . . them: - (H6, 8); common to the two of them: - (All H; L1).

8. on two fixed poles and a fixed axis: Among the Hebrew manuscripts, only H2 and H6, 8 follow the Arabic exactly. H4, 5 read: on two small poles and a fixed pole. H3, 7 read: on fixed poles and a fixed pole. These four manuscripts change the last letter of *qoṭer* (usually "diameter"; here, "axis") to read *qoṭev* (pole). The reading in H1 is similar to these variants, except for the last two words, where H1 reads *w ᵉ-qoṭev hataḥton* (the lower pole); axis: axes (H6,8).

9. axis: axes (H6,8). The Arabic also has the plural form.

[188] 1. solid: (L1).

2. whose shape: *tnu ᶜato* (whose motion) (H5).

3. whose shape. . . sphere: *spericum* (L1).

4. of that orb: + eccentric (H6, 8).

5. Its exterior surface: its two exterior surfaces (all H - H6, 8, with minor grammatical variants).

6. surface: - (H3); *mi-geshem* (of the body) (H4, 5), resulting in the redundancy *mi-geshem mi-gerem* (of the body of the body).

7. two surfaces: *superficie* (L1) = (H1).

8. sphere: + of the body (H6, 8).

9. equal: - (H2).

10. of the body: - (H2 - 5, 7).

11. with it: + with this motion (all H except H6, 8).

12. like a part: in a part (H6, 8).

[189] 1. When: H6 adds at the beginning of this passage: and when this orb moves from the west, as we said.

2. the eccentric orb: *ha-galgal ha- ᶜelyon ha-yoṣe ha-merkaz* (the upper eccentric orb) (H4, 5).

[190] 1. the plane of: - (L1).

2. parallel: - (L1).

3. two parallel circles: - (H6).

4. two surfaces: all H (save H6, 8) and L1 as well display this in the singular.

5. the center of which . . . world: - (H6, 8).

[191] 1. orb: *circulo* (L1); + *lo ha-m ᵉdummeh* (not the parecliptic) (H2).

2. is tangent: - (L1).

3. to the parecliptic orb: the smaller is: - (H2).

4. to it: - (H1). H8 adds: i.e., the parecliptic orb.

[192] 1. into nine: into seven (H6, 8).

2. parallel circles: surfaces (H1, 3 - 5, 7) = *ḥalaqim* (parts) (H2).

3. adjacent: which is below it (H6, 8).

[193] 1. two tangent points: all H save H6, 8 display this phrase in the singular.

2. orb: - (L1).

3. orb: - (L1).

4. the center of: - (H4, 5).

5. and the other. . . world: - (H6, 8).

6. For if. . . sun: - (L1); the orb of the sun: the eccentric orb (H6, 8).

7. the tangent point: *sh ^{e}tei n ^{e}qudot ha-dibbuq* (H6, 8) = *duos punctos contigionis* (L1).

8. the circumference of: - (all H - H6, 8).

9. orb: - (L1).

10. the remaining segment: *pars autem exiens in partem oppositam* (L1).

11. to its circumference: to the circumference of this orb (H6, 8).

[194] 1. and which. . . orb: - (L1).

2. That: + point (H6, 8); similarly, L1 adds: *punctum superius terminans* .

3. furthest distance: H1 reads here: *ha-merḥaq he-'arokh* . The usual term is *ha-merḥaq he-raḥoq* . Interestingly enough, H2 reads here: *ha-merḥaq he-'arokh w ^{e} -he-raḥoq* .

4. and that: and the point (H6, 8).

5. closest distance: *ha-merḥaq ha-qaṣar ha-qarov* (the close and near distance) (H2).

6. The point. . . sun: - (H4, 5, 7).

7. of the sun: - (H2, 4, 5, 7).

[195] 1. faces: - (H6).

2. orb: circle (H8).

3. 24 1/2 parts: *13 gradus et medietatem 1 gradus* (L1). Is this a scribal error?

4. preceding it: + towards the solstice (H6, 8).

5. his book known as: - (L1).

6. the *Almagest* : These is no uniform spelling in the Hebrew manuscripts. H1 and 2 read *magisṭi* ; H6 reads *magzisṭi* . H3 reads *magsa'ṭi* (?), the *gimel* written over the *somekh* , which may be the origin of the corrupt spelling *masa'ṭi* in H4 and 5. H8 has *magisṭo* .

[196] 1. of the ecliptic orb: of the ecliptic circle (H6, 8).

[197] 1. In H2 this sentence occurs after passage [198] with the following addition: and according to what Ptolemy said.

2. parts: + aforementioned (H6, 8).

[198] 1. and change position: - (L1).

[199] 1. the parts. . . part after part: point after point of the ecliptic circle (H6, 8).

2. in successive instants. . . points: *et singulis temporibus perveniunt ad singulis punctus* (L1)

3. it: the sun (H1).

4. the first point: the point which it had been facing (H6, 8).

5. has been completed: - (H3).

[200] 1. Thus the sun always: - (H2); always: - (H3, 5, 7).

2. to the circumference: in the plane (H6, 8).

3. orb: - (L1).

4. It: the sun (all H - H6, 8).

5. It then changes its position: *et cum separatur recedit ab auge* (L1).

6. nearer: + *ad terram* (L1).

7. reaches: + *ad punctum oppositum suo augi, scilicet* (L1).

8. of its orb: - (L1).

9. to ascend: - (H1, 3 - 5).

10. it returns to: - (L1).

11. Thus it is always: - (L1).

[201] 1. faces: + *describit* (L1).

2. arcs: *ḥalaqim* (parts) (H1).

3. it passes through: - (H2, 3, 7).

4. the parts of: - (H6, 8; L1).

5. in an [amount of] time. . . time: *diucius* (L1).

6. through: + the parts of (all H- H6, 8).

7. which is near. . . semicircle: - (H4, 5); the semicircle: - (H7).

[202] 1. its proper motion: *motus eius excentrialis* (?) (L1).

2. the motion. . . orb: its motion in the eccentric orb (H3).

3. that is, the equal motion: *et hic motus est equalis* (L1).

4. unequal: + *seu inuniformis* (L1).

[203] 1. place: *ha-yishshuvim* (the inhabited places). This Hebrew word is usually uses to translate the Arabic *masâkin* rather than *makân* , the word found in this place in the text. Our text states that this zone is uninhabitable.

2. on the earth: - (H6, 8; L1).

3. it: this line (H6, 8). The reference is to the "line of equality," i.e., the terrestrial equator.

[204] 1. in its perigee: *in oppositione sui augis* (L1).,

2. in a day and a night: + in every (H6, 8).

3. proximity: + *terre* (L1) = (all H).

4. a circle: + *bilti g ᵉdolah* ([which is] not great) (H6, 8).

5. to that circle: - (H2 - 5, 7) ;[closest] of the circles of the earth to this parallel circle on which the sun moves (H6, 8).

[205] 1. from the center of the world: from the center of the earth, that is, from the center of the world (H6).

2. in its perigee: *in oppositione sui augis* (L1).

3. also. . . earth: - (H2 - 4, 7).

4. we imagine. . . a great circle: *et ymaginati fuerimus lineam illam moveri cum puncto illius linee qui est in superficie terre, faciet circulum non magnum in superficie terre* (L1).

5. time: - (H6, 8).

[206] 1. conical: *m*^{e}*ḥubbar* (connected!) (H4, 5). This is probably a scribal error for *m*^{e}*ḥuddad* (conical). H8 has a lacuna here.

2. moves: + around the earth (H6, 8).

3. is not: *est* (!)(L1).

4. from its motion: - (H6, 8).

[207] 1. of the world: of the circle (H1); of the earth (H2- 5, 7).

2. the line. . . sun: - (L1).

3. the segment of it: this line (H6, 8).

4. perigee: *oppositum augis* (L1).

5. the length of the year: Various readings of this phrase occur in the Hebrew manuscripts . H2, 4, 5, and 7 add *b*^{e}*-khol* (in the whole), and thus their reading is similar to that of L1, which replaces "length" (Arabic *ṭûl*) with *totum* . H3 reads: on the earth for the whole year, though *ba-'areṣ* may be an error (of one letter) for *b*^{e}*-'orekh* (for the length).

6. between . . . earth: - (H2).

7. which we have mentioned: - (L1).

8. Because of this: *similiter* (L1).

9. on earth: - (all H save H6, 7; L1).

[208] 1. As one. . . hot: - (H7).

2. circle: + of the earth (H6).

3. to this path: *ad viam combustam* (L1).

4. It is said: *w ᵉ-yo'mar 'eḥad* (someone says) (H4, 5).

5. no living being: *nec ullum alia* (L1).

[209] 1. a diagram for: - (H2 - 5, 7).

2. according to: *r"l* [*r ᵉṣoni lo'mar*] (that is to say) (all H).

3. and for all. . . produces: - (H6, 8); produces: + by means of its motion (all H- H6, 8).

4. according to. . . is it: *scilicet circulorum qui fuint per motum solis* (L1).

Chapter Ten

The Orb of the Moon

[210] The orb of the moon is a round[1], spherical body, bounded by two parallel[2] surfaces whose center is the center of the world. The upper of the two surfaces is tangent to the concavity of the sphere of Mercury. The lower of the two touches the sphere[3] of fire. [211] The parecliptic orb, which we mentioned earlier, is in its upper surface. [212] This orb--that is, the orb of the moon[1]--moves in its entirety with a circular motion from east to west, contrary to the succession of the signs of the zodiac, on two fixed poles which face the poles of the ecliptic circle.[2]

[213] This orb is first[1] of all divided into two contiguous[2], round, spherical[3] sections whose center is the center of the world. One of the two encompasses[4] the other.[5] The concave surface of the upper section touches the convex surface of the lower section.[6] [214] The lower section[1], which is the one closer to the center of the world, moves also with a circular motion from east to west contrary to the succession of signs on two fixed poles other than the poles of the ecliptic orb. This orb is called the inclined [orb].

[215] A round[1] sphere is separated from this orb.[2] It is bounded by two round[3], spherical surfaces whose center is apart from the center of the world and outside of the plane of the parecliptic orb, and is equidistant from the poles of the inclined orb.[4] The upper[5] of its surfaces is tangent to the upper surface of the inclined orb in one point. Its lower surface is tangent to the lower of the surfaces of the inclined orb in one point also.[6] [215] This orb is called the eccentric orb.[1] It moves with a circular motion from west to east[2], according to the succession of signs, on fixed poles other than the poles which were previously [mentioned].

[217] A round sphere[1] is separated from this orb between its parallel surfaces whose center is equidistant from its poles, whose diameter is equal to the width of the eccentric orb, and whose

exterior[2] surface is tangent to its [the eccentric's] parallel[3] surfaces in two common[4] points. It is fixed in the body of this orb as a part of it.[5] It also moves with a circular motion about its own center on poles fixed for it. This sphere is called an epicyclic orb. [218] The eccentric orb is called the deferent orb because it is like a carrier* with respect to this sphere.

[219] The moon is a solid body with the form of a sphere, fixed in the body of the epicyclic orb. Its exterior[1] surface is tangent to the surface of the epicyclic orb in one point. [220] Thus, when the eccentric orb, which is the deferent of the epicyclic[1] orb, moves[2] on its poles, it sets in motion by means of its own motion the whole epicyclic orb in its entirety, because it [the epicycle] is like[3] a part of it. The body of the moon moves with it [the epicycle], because it is joined to it.[4]

[221] As the sphere moves with a circular motion, every point on it describes an imaginary circle. The center[1] of the epicyclic orb, by means of the motion of the eccentric orb, describes an imaginary circle whose center is the center of the eccentric orb.[2] This[3] is also called the eccentric orb. It is called the deferent orb because the center of the epicyclic orb moves on its circumference.

[222] If we imagine the plane of this circle to intersect the world, there then result in the surface[1] of the eccentric orb two parallel circles [that are] parallel as well to[2] the first circle, i.e., the middle one[3], whose center is the center of the eccentric orb.[4] A circle[5] is also produced in the surface of the epicyclic orb tangent to the two parallel[6] circles in two points, whose center--which is the center of the epicyclic orb--is on the circumference of the first circle. [223] The circle parallel to this circle which passes through the center of the moon is also called the epicyclic orb, and on it moves the center of the moon.

[224] Two circles are also produced in the surface of the inclined[1] orb. Their center is the center of the world and their poles are the poles of the inclined orb.[2] Also, there results in the surface of the first orb, which encompasses the inclined[3] orb, a circle parallel to the first two circles and intersecting the parecliptic orb.[4] [225] In the surface of the highest orb a circle which intersects the ecliptic

---

* The Arabic expression is *ka-l-ḥâmil* ; *al -ḥâmil* is the term for the deferent.

circle is also produced. This circle is also called the inclined orb of the moon.[1] [226] The inclination[1] of this circle from the ecliptic circle is that which is called[2] the latitude of the moon[3], because the center of the moon is always in the plane of this circle.[4]

[227] The two points of intersection of this circle and the ecliptic circle are called the nodes. The one which is the beginning of the northern half of the inclined circle is called the head, and the one which is the beginning of the southern half is called the tail. The inclination of this circle from the ecliptic circle is always the same unchanging inclination.

[228] If a straight line is drawn from the center of the world to the center of the deferent orb and extended rectilinearly in both directions to the circumference of the circle which is the eccentric orb[1], then the point which is the extremity of the line[2], near to the eccentric[3] center and furthest from the center of the world, is called the apogee of the moon, The opposite [point] is called the perigee of the moon.[4] [229] When the center of the epicycle is at its furthest distance[1], and we extend the line rectilinearly to the circumference of the epicyclic orb, then its [the line's] terminus[2] is also called the furthest distance of the epicyclic orb. The other extremity[3] of its diameter is called the shortest distance of the epicyclic orb.[4]

[230] The first of the moon's orbs moves with a circular motion on its poles, which face the poles the ecliptic circle, i.e., the poles of the parecliptic orb. By means of its own motion it sets in motion all of the orbs which are within it.[1] The orb moves with its motion against the order of the signs.[2] The entire place of the inclined circle moves against the order of the signs. This circle is the one of which we said that its circumference intersects the circumference of[3] the ecliptic circle. [231] Thus the two points of intersection move with this motion against the order of the signs. This motion is the one which is call the motion of the nodes.

[232] The second orb of the moon, the inclined orb, moves also with a circular[1] motion against the order of the signs on its two fixed poles, which are the poles of the inclined circle which is in its surface. It sets in motion by means of its own motion all that is within it. The eccentric orb thus moves with this motion against the order of the signs. The whole plane of the circle which is the eccentric orb moves with its motion[2], as do its center and its furthest distance.[3] The

motion of all of these[4] is against the order of the signs, about the center of the world, and in the plane of the inclined orb.

[233] The eccentric orb, which is [also] the deferent[1] orb[2], also moves with a circular motion[3] in the order of the signs on its two fixed poles. The epicyclic orb moves with its [the eccentric's] motion in the order of the signs. Its center moves[4] on the circumference of the eccentric orb.[5] [234][1] The body of the moon moves with the motion of the epicyclic orb in the order of the signs.[2]

[235] The epicyclic orb also moves with a circular motion on poles fixed for it, moving around its center.[1] The body of the moon moves with its motion. The center of the moon moves[2] on the circumference of the circle with is also call the epicyclic orb. [236] If it--that is[1], the center[2] of the moon[3]--is moving on the arc which is near the furthest distance of the epicyclic orb, then its motion will be against the order of the signs. If it is moving on the arc[4] which is near the closest distance, then its motion will be in order of the signs.

[237] As for the diameter of the epicyclic orb whose extremity is[1] the furthest distance that is on the rectilinear [extension] of the diameter[2] of the eccentric orb: when the epicyclic orb moves, and its center is displaced[3] from the point of furthest distance[4] of the eccentric[5] orb, and, along with it, the diameter whose two extremities are the furthest distance and the closest distance is [also] displaced, then this diameter does not continue to be in the situation in which it had been. [238] I mean that it does not always face the center of the eccentric orb, nor does it face the center of the world. That is, it it is extended rectilinearly, it does not end up at the center of the world nor at the center of the eccentric orb[1], but it [rather] will always face another point of that diameter near the closest proximity of the eccentric orb[2] whose distance from the center of the world is of a magnitude equal to the distance of the center of[3] the eccentric orb from it [the center of the world]. It always preserves this alignment* .[4]

[239] As for the center of the epicyclic orb and the point of furthest distance of the eccentric orb: their motions which we have

---

* In this passage Ibn al-Haytham describes the *prosneusis* of the epicyclic diameter and the determination of the "mean apogee." Cf. Neugebauer, *HAMA* , vol. 1, p. 68.

mentioned are opposed[1] to each other in two different directions.[2] One of them--the motion of the center of the epicyclic orb--is in the order of the signs. The other[3], which is the motion of the point of furthest distance, is against the order of the signs.[4] However, their distances from the mean position of the sun are always equal. In other words[5], the position[6] of the sun[7] is centrally located between these two points.

[240] The point of furthest distance moves around the center of the world. It moves from its facing the position of the sun until it returns to its facing [the sun] in a lunar month's time.[1] [241] The center of the epicyclic orb moves on the circumference of the eccentric orb. It leaves the point[1] of furthest distance and returns to it[2] twice in a lunar month's[3] time. At the moment of mean conjunction they all fall in line.[4] Thus the center[5] of the epicyclic orb and the point of furthest distance come together; then the two of them move [off] in different directions.

[242] When the center of the epicyclic orb has traversed half the circumference of the eccentric[1] orb[2], it arrives at the point of closest proximity[3] and is then in opposition to the point of furthest distance.[4] The point of furthest distance has [meanwhile] moved from its facing the sun[5], its distance from it [the sun] being [now] a quarter of a circle. Thus both of the points[6] are in quadrature with respect to the sun.[7] For this reason, the mean position of the moon[8] at quadrature is at it closest proximity to the eccentric orb.[9] [243] Thus, when the center[1] of the epicyclic orb moves [again], completing its circle, the point of furthest distance also moves [through] a quarter of a circle[2], and thus the two come together in opposition to the sun.[3] For this reason[4], the mean position of the moon[5] at the syzygies is at its furthest distance from the eccentric orb. [244] Thus the position[1] of the moon at conjunction and at opposition[2] is always at its furthest distance from the eccentric orb[3] and at the quadratures is at its closest proximity to it.[4]

[245] The body of the moon moves around the center of the epicyclic orb, sometimes[1] against the order of the signs, i.e., when it is in the upper part of the epicyclic orb, and sometimes in the order of the signs, i.e., when it is in the lower [part] of the epicyclic orb.[2] However, when it is moving against the order of the signs, it is not seen to retrograde, because the epicyclic orb[3] moves in the order of

the signs, and the body of[4] the moon moves with its motion [i.e., of the epicycle] which is faster[5] than the [particular] lunar motion which is opposed to it.[6] For this reason the motion of the moon is never[7] retrograde. Its motion in the upper part of the epicyclic orb is thus slower, whereas it is faster[8] in the lower [part] of the epicyclic orb.

[246] Six motions thus appertain to the moon: (1) the motion of the whole of its orb against the order of the signs, which is called the motion of the nodes; (2) the motion of the inclined orb, also against the order of the signs [but] on different poles, which is called the motion of the apogee; (3&4) the motion of the eccentric orb, which sets in motion the epicyclic orb in the order of the signs facing the center of the epicyclic orb and on the circumference of the inclined orb[1]--which[2] is called the motion in latitude.[3] This same motion, when it is taken with respect to its facing the ecliptic circle, becomes the motion in longitude; (5) the motion of the epicyclic orb, which sets in motion the body of the moon, around the center of the epicyclic orb,[4] which is called the motion in anomaly; (6) the motion of the diameter of the epicyclic orb whose extremity is the point of furthest distance[5] as it faces the assumed point; this motion is called the prosneusis.

[247] Let up imagine a straight line drawn from the center of the world to the center of the moon and extended rectilinearly. It will then reach to the circumference of the inclined orb, for the center of the moon is always in the plane[1] of the inclined orb.[2] Let us also imagine a great[3] circle drawn on the poles of the ecliptic orb, passing through the point on the circumference of the inclined orb at which the straight line terminated.[4] [248] This circle[1] will also intersect the circumference of[2] the ecliptic circle. Thus an arc of this circle will be produced between the ecliptic circle and the inclined orb. This arc is called the latitude of the moon. The point where this circle intersects the ecliptic circle[3] is the position of the moon on the ecliptic circle.

[249] When the moon moves with the motions which have been set forth, moving in the order of the signs by a combination of these motions[1], the straight line which we have imagined to pass through[2] its center also moves with these motions.[3] [250] The circle which passes through the poles of the ecliptic circle[1] also moves in the

order of the signs, as does its point of intersection on the circumference of the ecliptic circle.[2] The observed motion of the moon in longitude with respect to[3] the ecliptic is a result of this. [251] The motion of the straight line passing through the center of the moon with respect to the circumference of the inclined orb is the observed motion of the moon in latitude. By means of this motion, the arc which is between the two circles, which is the latitude, changes in magnitude. Sometimes it increases, and other times it decreases, and for this reason the latitude of the moon varies.[1]

[252] When the point of intersection of the ecliptic circle and the circle passing through its [the ecliptic's] poles, which [point] moves on the circumference of the ecliptic circle and is the moon's position in longitude[1], reaches the point facing the center of the sun, then it is the situation which is called the conjunction. When it [this point] reaches the point opposite the center of the sun, then it is the situation which is called the opposition.

[253] When the straight line which passes through the center of the moon and moves[1] in the plane of the inclined orb[2] reaches the point of intersection of the inclined[3] orb and the ecliptic circle, which is that which is called[4] the node[5], then, if the center of the sun happens to be facing that point[6], the center of the sun and the center of the moon [both] fall on this straight line. In this case there is a solar eclipse. [254] That is, the rays which go out from our eyes to the body of the moon are on that straight line or close to it. The body of the moon thus prevents them[1] from extending themselves rectilinearly so that they may reach the body of the sun. The body of the moon acts thus as an obstruction between our eyes and the body of the sun, concealing from us the light of the sun.

[255] The rays go out from [our] eyes* in the form of a cone whose apex is the point of vision[1] and whose base is the surface[2] of the visual object. If we look at the body of the moon at the moment of the eclipse, the rays which go out from our eyes are in the form of a cone whose apex is the point of vision[3] and whose base is the body

---

* This is the theory of extromission which was identified in medieval times with the view of the "mathematicians." Ibn al-Haytham in his *magnum opus* on the science of optics developed a version of the intromission theory. See A. I. Sabra, "Ibn al-Haytham," *Dictionary of Scientific Biography*, vol. 6 (New York, 1972)189-210, p.192.

of the moon. [256] Let us imagine the surface[1] of that cone to be extended rectilinearly in the direction of the sun. Then, if the entire[2] body of the sun falls in the interior of that cone[3], the entire sun is hidden from us.[4] In that case, it is said that the entire[5] sun is eclipsed. If that which falls inside the cone is part of the body of the sun, and the other part is outside[6] [the cone] , that part [inside] is hidden from us, and it is said that the [part] of the sun which is eclipsed is that part[7] which is inside the cone.[8]

[257] When the entire[1] sun is eclipsed, it does not tarry before it begins to be uncovered.[2] For when the cone of rays[3] which we have mentioned reaches the orb of the sun, its diameter[4] is equal to the diameter of the body of the sun.[5] Thus, when the moon begins to move after the fullness[6] [of the eclipse], and the cone [also] moves[7], the sun begins to be uncovered.

[258] If the straight line[1] reaches the point of intersection, and the center[2] of the sun happens to be facing the other point opposite it, then[3] the center of the sun, the center of the moon, and the center of the earth[4] will be on that straight line.[5] In this case, there is an eclipse of the moon. [259] That is [because] the moon does not have light [of its own], but rather it takes on light from the light[1] of the sun, since it is a glossy body.[2] When the sun stands opposite it, it receives its light; it then becomes illuminated by its light. That light[3] is reflected from its surface to the earth[4]; it [moonshine] thus illuminates it [the earth].[5]

[260] The earth is an opaque[1] body possessing[2] a shadow. It shadow is always opposite the position of the sun. The shape of this shadow is also a cone. This is [because] the diameter[3] of the sun is greater than the diameter of the earth, and the rays which go out from the circumference[4] of the body of the sun to the circumference of the body of the earth meet in the form of a cone, by whose surface the rays are circumscribed.[5] Within the conical surface there is an uninterrupted dark[6] shadow in which there is no light[7], [extending] from the body of the earth[8] to the point of concurrence, because the body of the earth blocks from it the light of the sun. This shadow[9] is that which is called night. [261] This cone goes beyond the orb of the moon. Thus, if the body of the sun, the body of the earth[1], and the body of the moon are on one straight line, and [if] the body of the earth is in the middle between the two[2] [other bodies], the body of

the moon will be in the shadow cone on account of what we have said, namely that the shadow is always opposite the body of the sun[3], and it reaches past the orb of the moon. Thus if it [the moon] is inside the shadow cone[4], the earth is an obstruction[5] between it and the sun and thus prevents it from receiving the light. It returns to its intrinsic nature, and it is seen at that time to be dark and without light.[6]

[262] When the three bodies are exactly[1] on the straight line, then the moon is in the middle of[2] this cone. If the moon is near that line[3] on which the two other bodies are [to be found], then the moon is in the side of that cone. The extent to which it is darkened[4] and the amount [of time] it stays[5] [in eclipse] depend on how close it is to that line. [263] If the entire moon is inside this cone-shaped shadow, all of it is darkened, and it is said that all of it is eclipsed.[1] If part of it is inside the cone and part of it is outside, then[2] part of it is darkened and part of it is illuminated, and it is said that part of it has been eclipsed.[3] [264] When[1] it is totally eclipsed and it is passing through the middle of the cone, its duration[2] [in eclipse] is long. If it is passing through the side of[3] the cone, its duration is short, because its exit from it will be quicker.

[265][1] As for [the fact] that sometimes the entire moon is illuminated, sometimes part of it is illuminated, and sometimes none of it is seen to be illuminated[2]: It is according to its position[3] with regard to the sun and [with regard to] us.[4] That [part] of the moon which is lit-up is a section which is close to half[5], and that part of it which our eyesight reaches is also[6] close to half.[7] The lit-up half is always the half opposite the sun and facing it[8], but this half[9] changes[10] its situation with respect to us. [266] If it [the moon] is in opposition[1], then the half which is facing the sun is the half which is near to us[2], because[3] in opposition it [the moon] is diametrically opposed to the sun[4], and we are in this case in the middle between the two. Thus a part of the moon is seen which is totally lit-up. [267] As it moves from opposition, our position is not medial between the two of them; it rather moves away from the middle position with respect to the two of them.[1] Thus the lit-up half is not[2] exactly the same as that facing us, but rather part of the half facing us[3] is from the lit-up part and part of it is from the dark part.[4] [268] The amount [of the lit-up part] which is opposite to us continues to

diminish[1] until it reaches conjunction. Then the lit-up half of the moon is[2] the upper half of the moon, which is the half near the sun. The half which is near us is totally[3] dark. [269] As it departs from the sun, it returns to its first situation, and the lit-up part is full.[1] The two circles--that is, the circle circumscribing the lit-up part[2] and the circle which circumscribes the dark part which is near us--intersect each other. At first[3] a small part[4] is visible to us. [270] This small part is circumscribed by two arcs.[1] One of them is the arc of the circle which bounds the part near us and the other[2] is the arc of the circle[3] which bounds the lit-up part.[4] Because this small part is circumscribed by two arcs and it is on the surface of a sphere[5], it is a crescent.[6] For this reason, that which is seen of the moon in the beginning and the end[7] of the month is a crescent.[8]

[271] It then continues to increase.[1] When the light reaches half of the part which is near us, the convexity of the arc which bounds the lit-up part faces us. Therefore the arc is seen[2] as a straight line.[3] The result is that what is seen of the lit-up part is encompassed by a straight line and half the circumference of the circle which bounds[4] the part near us. For this reason, the lit-up part of the moon which is seen in these instances is a semi-circle. It then continues to increase until it reaches it fullness.[5]

[272] This is the whole discussion on the orbs[1] of the moon and their situations.[2] In order that that which we have stated be clear, let us draw a diagram for the orb of the moon. In it we will list all the circles that we mentioned and the rest of the matters which we described.[3]

[210] 1. round: - (L1)

2. parallel: + circular (H4, 5, 7).

3. sphere: + concave (H2).

[212] 1. that is. . . moon: - (L1).

2. circle: orb (H1-3)

[213] 1. This orb is first: This first orb (H6).

2. contiguous: - (H1, 6, 8).

3. contiguous, round, spherical: - (L1).

4. encompasses: is in (H1- 4, 6).

5. one. . . other: - (L1).

6. lower section: There seems to be an ommision here in our Arabic text--we recall that this entire chapter is missing from Arabic K-- which was noticed by the Hebrew translators and/or copyists. H3 adds here: "It appears that the description of the first section is missing here. However, from what he says at the end [of the chapter] concerning the motions, [it is clear that] the first motion is that with which the orb moves in its entirety, and it is the motion of the first section." Ibn Paṭer seems to have been working from a more complete Arabic copy. H6,8 read here: "This upper section moves with a circular motion from east to west, against the order of the signs, on poles which face the poles of the ecliptic circle. By means of its own motion, it sets in motion all of the orbs of the moon which it encompasses. This is the motion which we mentioned previously in this chapter. This orb is called the orb of the

nodes (*galgal ha-t ^{e}li*)." I may add that the Alfonsine Latin translation displays a similar reading.

[214] 1. lower section: second section (all H - H6, 8) = second, lower section (H6, 8).

[215] 1. round: - (L1).

2. orb: + inclined (H6, 8).

3. round: - (H6, 8).

4. outside. . . orb: - (L1); and is. . . orb: - (H6, 8).

5. upper: lower (H2).

6. Its lower. . . also: - (H6, 8).

[216] 1. eccentric orb: + and this is the figure (H2). There is a blank space at this place in the manuscript where the figure should have been drawn.

2. west to the east: *ab oriente in occidente* (L1).

[217] 1. round sphere: *spera breuis* (L1).

2. exterior: visible (H6, 8; L1). The arabic *ẓâhir* means both "exterior" and "visible" ; hence variants such as the one noted here are quite common in our text.

3. parallel: - (L1).

4. common: - (H1; L1).

5. as a part: in a part (H3, 6 - 8). As in Arabic , the Hebrew prefix "*b* ^{e}-" means "in" and the prefix "*ka* " means "as"; however, the two letters are very similar in Hebrew, and hence this sort of variant is also quite common in our text.

[219] 1. exterior: visible (H6, 8).

2. in one point: - (H8).

[220] 1. epicyclic: + the center of (H1).

2. Thus. . . moves: - (H8).

3. like: in (H2 - 6).

4. to it: + as a part of it (H2, 4, 5) = + in a part of it (H1, 3, 7).

[221] 1. center: - (H8).

2. The center . . . orb: - (H6); whose center. . . orb: - (H8); orb: circle (H4, 5)

3. This: + circle (H8).

[222] 1. plane: two surfaces (all H).

2. parallel to : - (L1).

3. i.e., the middle one: - (all H and L).

4. two circles . . . orb: - (H4, 5).

5. circle: + great (H8). This is added above the line, ostensibly as a correction.

6. parallel: - (L1).

[224] 1. inclined: - (H1).

2. Two circles. . . orb: - (H6, 8).

3. inclined: - (H5)

4. the parecliptic orb: - (H3); orb: + *in duobus punctis* (L1).

5. In. . . intersects the: - (H3).

[225] 1. of the moon: - (H7).

[226] 1. inclination: + *distancia vel declinatio* (L1).

2. that which is called: - (L1).

3. of the moon: - (L1).

4. circle: - (H3) = orb (H1, 2, 8,).

[228] 1. circle. . . orb: *circuli excentrici* (L1); orb: - (H3, 7).

2. which is the eccentric. . . line: - (H4, 5).

3. eccentric: - (H6, 8).

4. The opposite. . . moon: - (H7).

[229] 1. furthest distance: *in auge* (L1).

2. its terminus: the other end of its diameter (H6, 8).

3. extremity: center (!) (H2 - 5, 7).

4. The other . . . orb: - (L1).

[230] 1. within it: + and it moves against the order of the signs (all H, with various wordings).

2. The orb. . . signs: - (H2 - 5, 7). H6, 8 read "the inclined orb" and then continue as in our text.

3. the circumference of: - (H1, 2, 7); circumference: + which is in the surface of the highest orb (all H, though H1 - 3 are missing "orb").

[232] 1. circular: - (H2).

2. The . . . motion: *et similiter circulus excentricus* (L1).

3. furthest distance: *aux* (L1).

4. of all of these: - (H2, 5, 7).

[233] 1. deferent: inclined (H4, 5). The two words in Hebrew--*noṭeh* (inclined) and *nose* (deferent)--are orthographically similar.

2. The. . . orb: *tertius autem orbium lune scilicet orbis excentricus* (L1).

3. motion: + great (!) (H4, 5).

4. It center moves: Its center is embedded (H2 - 5, 7; H2 does have "moves" as well).

5. Its center. . . orb: - (L1).

[234] 1. L1 is missing this passage.

2. of the signs: - (H4, 5).

[235] 1. its center: it (H6, 8).

2. with its motion. The center of the moon moves: - (H6, 8).

[236] 1. it--that is: - (H1).

2. center: + *ha-*^{c}*elyon* (higher or highest) (!) (H3 - 5, 7).

3. moon: + the orb of (H8).

4. which is . . . arc: - (H2, 5, 7).

[237] 1. whose extemity is: which is tangent to (H2 - 5, 7).

2. diameter: pole (H3). The Hebrew words *qoṭer* (diameter) and *qoṭev* (pole) are orthographically similar.

3. when. . . displaced: - (H4, 5).

4. point of furthest distance: *puncto augis* (L1).

5. eccentric: + deferent (H2) = + deferent of the distance (*ha-nose ha-merḥaq*) (!) (H3 - 5, 7).

[238] 1. That is. . . orb: - (H8).

2. That is. . . eccentric orb: *set* (sic) *erit mundi directe respiciens alium punctum illius orbis versus hadid oppositum augi excentri* (L1). The meaning here of *mundi* is unclear.

3. the center of: - (H4-8).

4. It always. . . alignement: - (L1); alignement: + and it returns (*yashuv* or *yashshiv*) with it with this motion the circumference of the epicyclic orb (all H). H6, 8 add *ha-qoṭer* (the diameter) as the subject of this sentence; they also exhibit Ibn Paṭer's term for "to return" (in the transitive sense), which is *yaḥzir* .

[239] 1. opposed: small (!) (H4, 5).

2. in two different directions: - (L1).

3. the motion. . . The other: - (H4, 5).

4. is against. . . signs: -(H4, 5).

5. the mean. . . words: - (H2 - 5, 7).

6. the position: + *medius* (L1).

7. In other. . . sun: - (H6, 8).

[240] 1. in a lunar month's time: (H2 -5,7); month's: + mean (H1, 6,8).

[241] 1. leaves the point: moves from the motion (!) (H8).

2. The center. . . to it: - (H2, 5, 7).

3. month's: + mean (H1, 6, 8) = *medi* (L1).

4. they all fall in line: - (L1).

5. center: - (H6, 8).

[242] 1. eccentric: - (H2).

2. the circumference. . . orb: the circle (H6, 8).

3. it arrives. . . proximity: - (L1); proximity: distance (all H - H6, 8).

4. and is then. . . distance: - (H2 - 5, 7).

5. sun: + position of (H6, 8).

6. both of the points: all the points (H6, 8).

7. to the sun: + *loci* (L1).

8. of the moon: - (H7).

9. at quadrature is at its closest proximity of the eccentric orb: *et ideo erit luna secundum mensem cursum in quadratis in oppositio augis excentrici* (L1).

[243] 1. the center: + the point of (H6, 8).

2. a quarter of a circle: through an amount such that its motion, along with the motion of the sun, will complete half of the circle (all H).

3. sun: + the position of (all H - H4, 7).

4. the sun. . .For this reason: - (H4).

5. moon: sun (H6, 8). H6, 8 add *b ᵉ-emṣa ᶜ*, which should be *ha-emṣa ᶜi* (mean).

[244] 1. position: + mean (H6, 8).

2. at conjunction and at opposition: *in coniunctionibus magnis* (L1).

3. Thus the position. . . orb: (H2, 3).

4. at its closest proximity to it: *in oppositione augis* (L1).

[245] 1. sometimes: - (all H - H6, 8).

2. and sometimes. . . epicyclic orb: - (all H - H1).

3. epicyclic orb: center of the orb (H4,5,7).

4. the body of: - (H6, 8).

5. faster: + always (H6, 8) = + *sempter* (L1).

6. which is opposed to it: in the epicyclic orb (H2 - 5, 7).

7. is never: + *videtur* (L1).

8. faster: - (H6, 8).

[246] 1. which sets. . . inclined orb: *deferentis orbis declivis* (L1).

2. in the order. . . which: and it is the motion at (*bi* -) the center of the epicyclic orb facing the inclined orb. The motion of the inclined orb... (H6, 8).

3. in the order. . . latitude: - (H3).

4. around. . . orb: - (H6, 8).

5. whose extremity. . . distance: - (L1).

[247] 1. plane: $t^{e}nu^{c}ah$ (motion) (!) (H6).

2. for the center . . . orb: - (H2 - 5, 7, 8; L1).

3. great: - (H6, 8).

4. at which. . . terminated: - (H2 - 5, 7).

[248] 1. This circle: The half of this circle which passes through the end of the line (H6, 8).

2. the circumference of: - (L1).

3. This circle. . . circle: - (H2 - 5, 7).

[249] 1. by a combination. . . motions: - (L1).

2. pass through: face (H6).

3. with these motions: - (L1).

[250] 1. circle: orb (H8).

2. as does. . . circle: - (H1); H6, 8, which, alone among the Hebrew manuscripts contain this phrase, have some minor textual problems here.

3. with respect to: H1 adds here *b* *e* *-hiṣṭaref* , which can be translated as "with the combination of" or "with the addition of." Was the translator adding a second possible translation of the Arabic *'iḍâfah* ?

[251] 1. in magnitude. . . varies: *secundum magus et minus* (L1); varies: diminishes (H2 - 5, 7).

[252] 1. of the ecliptic. . . longitude: *per motum illius circuli* (L1).

[253] 1. moves: - (H1, 6, 8).

2. plane of the inclined orb: plane of the orb in the middle (*b* *e*-*emṣa* *c*) (H1).

3. inclined: - (H1, 6, 8).

4. the point. . . which is called: - (L1).

5. node: the head (all H - H6, 8); H1, 4, 5 add: of the dragon.

6. then, if the center. . . point: - (H6, 8).

[254] 1. are on. . . prevents them: - (H2).

[255] 1. vision: + from our eyes (H6).

2. surface: - (H1, 2, 7).

3. and whose base. . . vision and: (all H - H6, 8).

[256] 1. the surface: - (H6).

2. entire: - (L1).

3. to be extended. . . cone: - (H7).

4. from us: *me-he-ra'ot* (from being seen) (H6, 8).

5. entire: + body of the (H8).

6. and the other part is outside: - (L1).

7. it is said. . . that part: - (H2).

8. it is said. . . cone: the eclipse is not total. When the entire body of the sun falls inside of the cone, they say. . . (H6, 8). This leads into the following passage; that part. . . cone: *quod eclipsis particularis* (L1).

[257] 1. entire: + body of the (H8).

2. it does not. . . uncovered: *nullam moram facit* (L1); to be uncovered: H6, 8 correctly translate the Arabic, adding "to shine." The other Hebrew manusacripts use the puzzling expression *l e - ha'ir b e-halikhah*, literally "to shine in [its ?] movement." The source of this confusion is the multiple meanings of forms of the Arabic root *s.f.r.* . The form found in our text, *isfâr*, means "unveiling" or "uncovering." However, other forms of this root do in fact mean "to travel", "to go," etc.

3. rays: of vision (H6, 8).

4. diameter: of the surface of the base (all H).

5. body of the sun: + when the moon in in the furthest distance (all H) = *quando est in auge* (L1).

6. when. . . fullness: the moon moves (*yani ᶜa* ; the word may possibly be read *yagi ᶜa* , to reach) in motion after conjunction (all except H6, 8); H1 reads "another motion" instead of "motion after". H1 here uses Ibn Paṭer's term for conjunction, *ḥibbur* ; after the fullness: - (L1).

7. moves: + in that instant (*sha ᶜah*) (H6, 8).

[258] 1. line: + which passes through the center of the moon (H6, 8).

2. the center of: - (H6, 8; L2).

3. the center. . . then: - (H2).

4. and the center of the earth: - (H6, 8).

5. straight line: diameter (H2).

[259] 1. the light of: - (all H - H3 - 5).

2. since it is a glossy body: - (all H and all L).

3. That light: + which it received (H6, 8).

4. earth: + the surface of (H6, 8).

5. that. . . illuminates it: - (L1); it thus illuminates it: - (H6, 8).

[260] 1. opaque: large (H4, 5).

2. possessing: *faciens* (L1).

3. diameter: - (H1, 2, 4, 7) = body (H3, 5).

4. circumference: - (H6, 7) = center (H5).

5. by whose. . . circumscribed: - (H7; L1).

6. dark: + ᶜ*avah* (thick) (H6, 8).

7. in which there is no light: - (L1).

8. meet. . . earth: H3 is missing part of this phrase.

9. shadow: - (H2).

[261] 1. earth: *shamayim* (heavens) (!) (H4, 5).

2. between the two: - (L1).

3. of the sun: - (H4, 5).

4. cone. . . shadow cone: - (L1).

5. obstruction: + intervening (H2).

6. and without light: - (L1).

[262] 1. exactly: - (L1).

2. the middle of: - (H6, 8).

3. line: *ṣad* (side) (H2).

4. darkened: *y ᵉḥattekh* (H6, 8). The root of this word means to intersect, but its form and meaning here are unclear.

5. stays: + in darkness (H6, 8).

[263] 1. it is said. . . eclipsed: - (L1).

2. part . . . then: - (H2, 5, 7).

3. part of it is darkened. . . eclipsed: *erit eclipsis particularis* (L1); - (H5).

[264] 1. When: - (H5).

2. duration: + *b e-afelah* (in darkness) (H2 - 5, 7).

3. the side of: - (L1).

[265] 1. All H except H6, 8 add a new chapter chaper title: On Light and Darkness in the Moon.

2. As for. . . illuminated: Various versions of this sentence are found in the Hebrew manuscripts. H2 - 5, 7 read: as for the fact that the moon is sometimes seen illuminated and sometimes dark (H4, 5 add: [as a] dark body); H1 exhibits a similar reading, adding: No illuminated part of it is seen. H6, 8 replace this sentence with: As for the fact that sometimes the entire moon is illuminated, sometimes none of it is seen, and sometimes half of it seen, or more or less--all these visible appearances (*mar'ot nir'ot*) of it...; sometimes. . . illuminated: *quare vero in quibusdam temporibus videtur luna illuminata et in quibusdam obscura* (L1).

3. its position: + *diversorum situum* (L1).

4. and us: - (all H - H6, 8).

5. to half: - (H6, 8).

6. also: *semper* (L1).

7. close to half: less than half (H6, 8); the added word *be-qaṭnut* in the other Hebrew manuscripts may give the same meaning.

8. and facing it: - (L1).

9. half: + lit-up (H8).

10. changes: + *semper* (L1).

[266] 1. If it. . . opposition: At the time of full moon, which is the time of opposition (H6, 8).

2. near us: seen to us (H6, 8).

3. If it. . . because: - (L1).

4. is diametrically opposed to the sun: All H save H6, 8 have unintelligible wordings. Ibn Paṭer's translation (H6,8) reads here: "because at opposition [H8 adds: the position of the sun !] it is inclined from the ecliptic circle slightly, since its motion is always in the plane of the inclined circle. Thus it is slightly inclined from true opposition. For this reason, it [the moon] is not hidden from the sun at the moment of opposition and it avoids being eclipsed, because it is not in the node, and so it is illuminated by the light of the sun." This explanation why lunar eclipses do not occur every month is rather common to medieval treatises, and its inclusion here may offer further evidence that Ibn Paṭer was working with a better Arabic manuscript than those available to us today.

[267] 1. the two of them: us (H4, 7).

2. is not: H1 - 5 read "is"

3. it rather. . . facing us: it rather, for example, is inclined from between them (H6, 8).

4. and part of it. . . part: - (H1).

[268] 1. continues to diminish: - (H1).

2. the lit-up. . . is: (H7).

3. totally: - (L1).

[269] 1. and the lit-up part is full: - (L1); full: - (H2 - 5) = inclined (H6, 8) = *m ^{e}qabel* (receives) (H1).

2. the two. . . part: - (H2, 5, 7).

3. At first: - (L1).

4. part: + of the lit-up part (H6, 8).

[270] 1. arcs: circles (H6, 8).

2. is the arc. . . and the other: - (L1, H2).

3. which bounds. . .circle: - (H2).

4. and the other. . . lit-up part: - (H6, 8).

5. the surface of the sphere: *superficiem terre spere* (L1).

6. it is a crescent: - (L1; there is a lacuna in the manuscript); and the surface of the sphere is seen from afar as a flat (*pashuṭ*) plane; this part is always seen as a small crescent (H6, 8). The other Hebrew manuscripts read: it is called a *al-hilâl* ; H1, 2 add correctly "in Arabic"; the transliteration varies in the different manuscripts. The word appearing in H6, 8 is *yeraḥ* , which is another Hebrew word for moon that Ibn Paṭer seems to have set aside for "crescent".

7. in the beginning. . . end: in the middle (*ḥeṣi*) (sic in H8). All the other H lack "and the end."

8. crescent: *ki-d ^{e}mut ta ^{c}ar* (like the appearance of a razor) (All H save H6, 8).

[271] 1. It. . . increase: The lit-up part continues to become distant from the sun (H6, 8).

2. seen: + from afar (H6, 8).

3. line: + and the surface of the sphere is seen to be flat (H6,8).

4. bounds: + the arc which bounds (H2 - 5, slightly different in H1).

5. reaches its fullness: returns to the likeness of that which it was (H1 - 5, 7).

[272] 1. orbs: orb (all H).

2. and their situations: - (H3, 6 -8; L1).

3. In it . . . described: - (L1). H2 supplies the following caption to a figure, which, however, is not found in the manuscript: The large cone is the cone of the rays from the sun which surrounds the sphere of the earth. Its black part is the earth's shadow. The small cone within it is the cone of visual rays which goes out to the body of the moon and reaches to the body of the sun. The lines going out from the body of the sun to the body of the moon in the quadrants are the rays of the sun going out to the body of the moon, from which derive the light of [both] the sun and the moon.

Chapter Eleven

The Orb of Mercury

[273] The orb[1] of Mercury is a round, spherical body bounded by two parallel[2] spherical surfaces whose center is the center or the world. The upper of the two surfaces[3] is tangent to the concavity of the orb of Venus. The lower of the two surfaces is tangent to the convexity of the orb of[4] the moon. [274][1] The parecliptic orb, which is produced by the motion of the orb of[2] the sun, is in the upper surface. [275] The orb--that is, the orb of Mercury[1]--moves slowly with a circular motion around the center of the ecliptic orb and on its [ecliptic's] poles from west to east[2] in the order of the signs.

[276] Separated from this sphere is a round[1] sphere[2] that is bounded and set apart[3] by two parallel spherical[4] surfaces whose center, which is [also] the center of the sphere, is removed from[5] the center of the world and from the plane of[6] the parecliptic orb.[7] Its upper[8] surface is tangent to the upper surface of the first orb in one common[9] point. It moves on fixed poles other than the poles of the ecliptic orb about its own eccenter with a circular motion against the order of the signs. This orb is called the dirigent* orb of the deferent orb.

[277] Another round[1] sphere is separated from[2] this orb.[3] It is also bounded by two parallel[4] spherical surfaces whose center, which is the center of the sphere[5], is removed[6] from the center of the world, from the center also of the dirigent orb and from the plane of the parecliptic orb, in the direction in which the center of the dirigent orb[7] [is removed]. Its distance from the center of the dirigent [orb] is[8] half the distance between the center of the dirigent

---

* I owe this technical term to my colleague F. Jamil Ragep. The Arabic word is *al-mudīr* . For detailed analyses of Mercury's complicated theory, see W. Hartner, "The Mercury Horoscope of Marcantino Michiel of Venice," *Vistas in Astronomy* 1 (1955) 84-138, and O. Gingerich, "The Mercury Theory from Antiquity to Kepler," *Actes [du] XIIe Congrès International d'Histoire des Sciences* , vol. 3a, *Science et Philosophie Antiquité-Moyen Age-Renaissance* (1968) 57-64.

[orb] and the center of the world.[9] Its upper surface is tangent to the upper surface[10] of the dirigent orb[11] in one point. This eccentric orb is called the deferent orb.[12] This orb also moves on poles fixed for it other than[13] the first poles with a circular motion[14] in the order of the signs.

[278] Another round sphere is separated from this sphere also between its parallel[1] surfaces. Its center is equidistant from its fixed[2] poles. Its exterior[3] surface is tangent to the parallel[4] surfaces in two common[5] points, and its diameter[6] is equal to the distance between the parallel[7] surfaces.[8] This sphere[9] is called the epicyclic orb. This sphere also moves on two poles fixed for it with a circular motion about its center. [279] The body of[1] the star, that is[2], Mercury, is a solid body of spherical form fixed in the body of the epicyclic orb. Its exterior surface touches the surface of[3] the epicyclic[4] orb[5] in one point.

[280] When the deferent orb moves on its own special[1] poles, it sets in motion by means of its own motion the entire[2] epicyclic orb, and it carries it by means of its own motion about its [deferent's] center.[3] An imaginary[4] circle is produced from the motion of its [epicycle's] center[5] whose center is the center of the deferent orb. It[6] [this circle] is not in the plane of the parecliptic orb. [281] If we imagine this plane to intersect the world, it will produce in the two parallel[1] surfaces of the deferent orb two parallel circles which are parallel to the first circle. Their center is the center of the deferent[2] orb.[3]

[282] Two parallel circles are also produced in the surface of the first orb[1] whose center is the center of the world.[2] Their center is the center of the world. The larger of the two, which is the one in the upper[3] surface[4], intersects the parecliptic orb in two opposite points. This circle[5] is called the inclined orb.

[283] A circle is also produced in the surface of the highest orb which intersects the ecliptic circle in two opposite points which are called the nodes. This circle[1] is also called the inclined orb.[2]

[284] Let us imagine a straight line which is drawn from the center of the world to the center of the dirigent orb and which reaches to its circumference in both directions. Let the center of the deferent[1] orb also lie on it in the direction of the extremity of the line[2] in which lies the center of the dirigent orb. The extremity of

this line which is in the direction in which the two centers lie is the point of tangency with the surface of the dirigent orb.[3] It is the furthest distance[4] of the eccentric orb, and it is also the point of the orb of Mercury[5] furthest from the center of the world.[6] Opposite it is the closest distance[7] to it* .

[285] The first of Mercury's orbs, which encompasses all of its orbs[1], moves with a circular[2] motion about the center of the world on poles which face the poles of the ecliptic orb. It carries all the orbs of this star[3] that it encompasses. It carries them[4], the centers of the deferent orb and the dirigent orb, the furthest distance[5] and the closest distance[6], and the two points of intersection.[7]

[286] This motion is a slow motion[1] in the order of the signs. Its period[2], according to what Ptolemy mentioned in the *Almagest* [3], is one part of the parts of the circle[4], which are 360, in each hundred years.*

[287] The second of Mercury's[1] orbs, which is the dirigent orb, moves about its center on fixed[2] poles. It sets in motion the entire deferent[3] orb, its center, and all that is in it[4] with a circular motion against the order of the signs around the center of the dirigent orb. [288] The deferent orb also moves with a circular motion about its center and on its own special[1] poles from west to east. It moves the epicyclic orb by means of its own motion. [289] It moves its

---

* Ibn al-Haytham here refers to the point diametrically opposite the apogee which, however, is not the perigee, that is, it is not the point of minimum earth-planet distance. Observations led Ptolemy to conclude that in fact there are two points of minimum distance , about 120° distant from and on opposite sides of the apogee. See Neugebauer, *HAMA* , vol. 1, p. 159, and W. Hartner, "Medieval Views on Cosmic Dimensions and Ptolemy's *Kitâb al-Manshûrât* ," *Mélanges Alexandre Koyré* (Paris,1968) 254-82, p. 267, n. 5. This particular issue does not really have any bearing on the primary objective of our treatise, namely the pairing of motions to orbs, and Ibn al-Haytham consequently ignores it. Nevertheless, one would think that the notion that a planet approaches the earth twice in each revolution would be cause for some cosmological musing. Some cosmological peculiarities of Mercury's model are discussed by W. Hartner, "The Mercury Horoscope of Marcantonio Michiel of Venice," *Vistas in Astronomy* , 1(1955) 84-138, pp. 125 ff, esp. p. 131. However, we do not find convincing Hartner's evidence for the influence of our text upon Peurbach; in the manuscripts of *On the Configuration* that we have examined there are no figures of the sort which interest Hartner.

* Book 9, ch. 7; G. J. Toomer, *Ptolemy's Almagest* , p. 453.

[epicycle's] center on the circumference of[1] the circle which is also called the deferent orb with a uniform motion in the order of the signs. [290][1] The body of the star also moves in the order of the signs along with the whole[2] epicyclic orb.

[291] The epicyclic orb also moves with a circular motion about its center, and its sets in motion[1] the body of the star[2], i.e., Mercury, with a circular motion.[3] [292] An imaginary[1] circle is produced from the motion of its center[2] whose center is the center of the epicyclic orb.[3] It is also called the epicyclic orb.[4] This circle is inclined[5] to the plane of the eccentric orb.[6] [293] The center of the star[1] always[2] moves on the circumference of this circle. When it [Mercury] is in its [epicycle's] higher [part], its motion is in the order of the signs and when it is in its lower [part], it moves against the order of the signs.[3]

[294] When the deferent orb moves, it sets in motion the entire epicyclic orb. Its diameter[1] which is on the rectilinear [extension] of the line on which lie the centers of the orbs also moves, and its position changes.

[295] Let us imagine a point on the middle of the line which [runs] between the center of the world and the center of the dirigent orb, and[1] let us imagine the diameter of the epicyclic orb, after its separation from the furthest distance[2], to be extended rectilinearly until it reaches this point. Then this line is called the dirigent line of the diameter of the epicyclic orb. For the epicyclic orb[3] always moves on the circumference of the deferent orb, and its above-mentioned diameter is on the rectilinear [extension] of the line which is drawn from its center to the above -mentioned point. The circle which is described by the extremity of this line is also called the eccentric orb.[4] [296] The three centers lie in the same direction from the center of the world, which is that [direction] near the furthest distance.[1] The first of them is the dirigent center of the diameter of the epicyclic orb. Next to it is [the center of] the dirigent [orb] of the deferent[2] orb, and next to it is the center of the deferent orb* . The distances between them are equal. [297] When the dirigent orb of the deferent orb moves, it sets in motion the center

* The situation is described at apogee, i.e. when the epicyclic center is at its furthest distance from the center of the world.

of the deferent orb.[1] It circles around[2] in its motion until it coincides with the dirigent center[3] of the epicyclic diameter. It then separates from it, and in its return it ends up where it was at first.[4]

[298] The motions of the deferent orb and its dirigent orb are opposed to each other, as we said earlier.[1] However, the dirigent orb moves[2] against[3] the order of the signs. It makes one revolution in a solar[4] year. Its diameter which passes through the center of the deferent orb, which turns it[5], comes back to coincide[6] with the first line on which lie the furthest distance and the closest distance,[7] and on which lie also the centers of the orbs, once in each year, also according to the first description[8] [given in passage 297].

[299] The deferent orb, which turns[1] the epicyclic orb, moves in the order of[2] the signs.[3] It makes two revolutions in this period of time, i.e.[4], a solar year, so that the line turning the diameter of the epicyclic orb will coincide[5] with the line that runs between the center of the deferent orb[6] and the center of its dirigent orb twice in this year.[7] [300] This coincidence with the line on which lie the centers of the orbs [happens] once near the furthest distance[1] and once near the closest distance,[2] similar to the situation with the orb of the moon.[3] [301] As a result of this motion, the center of the epicyclic orb[1] moves from facing a point on the inclined circle until it returns to it in [a period of] one solar[2] year. It has [in this time] moved twice along the circumference of the deferent orb.

[302] It is also said[1] to be always[2] facing the mean position of the sun. This means that the body of the star is found by continued observation to keep equal distances from the mean position of the sun--whether ahead of it or behind* --in its direct motion with regard to the ecliptic orb and [in] its retrograde motion.

[303] This motion, i.e., retrograde and direct, is that which is proper to the epicyclic orb. [304] The motion[1] of the star in the epicyclic orb[2] also keeps equal distances in its retrograde and direct

---

* "Ahead" and "behind" refer to Mercury's situation as a morning or evening star, respectively. Ptolemy declared that the maximum elongations to the east and to the west are equal, but some extrapolation from the observational data was necessary, as he himself admits. See Neugebauer, *HAMA* , vol. 1, pp. 159-161, and O. Gingerich, "The Mercury Theory from Antiquity to Kepler," *Actes du XIIe Congrès International d'Histoire des Sciences* ,IIIa (Paris,1968) 57-64, pp. 57-59.

[motion]--that is, in the upper and[3] lower [parts] of its epicyclic orb--only[4] from[5] the very[6] diameter which determines by its extremities the midpoint[7] of the retrograde and direct [motions]. This diameter, which is the one passing through the dirigent line[8], always faces[9] the position of the sun. For this reason, the center of the epicyclic orb always faces the position of the sun.[10]

[305] The body of the star moves about the center of the epicyclic orb. It faces in its motion, as shall explain, parts of the ecliptic circle.[1] When it is in the upper [part] of its epicyclic orb, it moves in the order of[2] the signs, and when it is in the lower [part], it moves against the order of the signs, contrary to the situation of the moon. However, given this[3], its motion outstrips the equal motion of its epicyclic orb[4] which is in the order of[5] the signs. [306] The situation which is called[1] the retrograde [motion] results, due to the difference of the two motions. That is, when it is the lower [part] of its epicyclic orb, its motion in this position is against[2] the order of the signs. The whole[3] epicyclic orb [participates] in the motion of the deferent orb in the order of the signs. Its [the deferent's] motion[4] outstrips that of the epicyclic orb.[5] That is, when it moves, passing over an arc of the ecliptic circle whose measure is, for example, one part retrograde, the epicyclic orb, by means of the accidental[6] motion of the deferent orb, draws it back by an amount of one half on one part. Thus it[7] appears to have retrograded on an arc of the ecliptic circle whose measure[8] is one half of one part. Thus the net [motion] is retrograde. As long as it fits this description[9], it is said to be retrograde.[10] [307][1] When its motion is in the upper [part] of the epicyclic orb, which [motion] is in the order of the signs, the two[2] motions are both in the order of the signs.[3] They join in one motion, and it is said at that time to be direct.

[308] The inclination of the inclined orb is small. However, it does not always [keep] its position but ,instead, it changes, in so far as the whole plane moves[1], approaching the plane of the ecliptic, and it continues [moving] in this manner until it coincides with the plane of the ecliptic.[2] Then it separates from it[3] and inclines in the other direction until it reaches a like amount of inclination, and then it returns. So it is always.

[309] The diameter of the epicyclic orb whose endpoints are the furthest distance and the closest distance[1] of the epicyclic orb[2] is not

always[3] in the plane of the eccentric orb, nor does it always keep to one position. Rather, it moves[4] on the circumference of[5] a small circle, as if it [the circle] were at its [diameter's] extremity.[6] It drives with it the plane of the epicyclic orb.. This circle is perpendicular to the plane of the eccentric orb, and its center is in it. The motion of[7] this diameter about this circle is called the inclination of the epicyclic orb.[8]

[310] Also, the diameter which intersects this diameter at right angles does not keep to one position, but it moves on a small circle as if it [the circle] were[1] at its [diameter's] extremity. It [circle] is perpendicular to the plane of the ecliptic circle.[2] This motion is called the slant of the epicyclic orb. Its speed[3] is equal to the first motion. The passage of the star in latitude takes place according to these two motions and the motion of the inclined orb* .[4]

[311] When the epicyclic orb moves on the circumference of the eccentric[1] orb, moving from the furthest distance[2] and going towards the point of intersection, the inclined orb begins its motion toward the ecliptic orb. When the epicyclic orb reaches the point of

---

* Ptolemy's account of the motion in latitude of the inferior planets is quite involved, to the point that he himself felt obliged to add to his technical presentation a short dissertation on the concept of simplicity. See Neugebauer, *HAMA* , vol. pp. 217 ff. An excellent summary of Ptolemy's latitude theory is found in J. L. E. Dreyer, *A History of Planetary Theory from Thales to Kepler* (rprt. New York,1953), pp. 198-200; for the inferior planets especially see the detailed account and clear drawings in N. Swerdlow, "The Derivation and First Draft of Copernicus' Planetary Theory: A Translation of the *Commentariolus* with Commentary," *Symposium on Copernicus. Proceedings of the American Philosophical Society* , vol. 116, no. 6 (Philadelphia,1973) 423-512, pp. 494-96.

Mercury's motion in latitude has three components: an oscillation of the eccentric inclined orb about the ecliptic; a motion of the epicyclic apsidal diameter, i.e. the diameter which runs through the closest and furthest distances of the epicycle from the center of the world; and a motion of the diameter orthogonal to the one just described. Each of these last two is pictured as a motion of the extremity of the epicyclic diameter about a circle that is imagined to be upon the epicyclic sphere. The motion of the apsidal diameter is called the inclination (*mayl*), and it controls the angle made by the plane of the epicycle with the plane of the eccentric; hence its small circle is centered upon and normal to the plane of the eccentric. The motion of the other diameter is called the slant (*'inḥirâf*), and it controls the angle made by the plane of the epicycle with the ecliptic; its circle is centered upon and normal to the plane of the ecliptic. See also the following notes.

intersection[3], the inclined orb coincides with the plane of the ecliptic orb. [312] When the epicyclic orb moves in the direction of the closest distance[1], the inclined orb also moves. It separates from the plane of the ecliptic orb, taking the furthest distance[2] in the other direction, which, in this case, is the northerly direction.[3] The closest distance[4] moves in the southerly direction. When the epicyclic orb arrives at the closest distance[5] of the eccentric orb, the inclination of the inclined orb has reached its limit[6].

[313] When also the epicyclic orb moves in the direction of the other of the two points of intersection[1], the inclined orb also moves, returning in the direction of the plane of the ecliptic orb. When it reaches the point of intersection, the inclined orb also coincides with the ecliptic orb.[2] [314] When it moves from the point of intersection[1] in the direction of the furthest distance[2], the inclined orb also separates from the plane of the ecliptic orb, returning towards the first direction, which is the south. However, the center of the epicyclic orb is always[3] inclined from the epicyclic circle in the southerly direction only* .

[315] The inclination of the epicyclic orb from the plane of the eccentric inclined orb is [accounted for by] the motion of its diameter about the small circle. It begins when the epicyclic orb is in the furthest distance[1] of the eccentric orb. This diameter leaves the plane of[2] the eccentric orb, moving toward the north. It drives[3] with it the plane of the epicyclic orb. Then it returns again to the plane of the eccentric orb. It then leaves it[4] [heading] to the south, and it then also returns to it. So it is always.[5]

[316] The diameter which intersects this diameter begins to depart [from the plane of the ecliptic] when the[1] epicyclic orb is in the point of intersection, which is the head or the node.[2] It moves toward the north on the circumference of the small circle which is perpendicular to the ecliptic circle.[3]

---

* The presentation here seems to follow that of the *Almagest* XII, 3-4; cf. Toomer, *Ptolemy's Almagest* , p. 601 . In the *Planetary Hypotheses* , one of the varying inclinations--either that of the plane of the inclined orb or that of the plane of the epicycle--is taken to be fixed, though it is not at all clear which one. See Neugebauer, *HAMA* , vol. 2, p. 909, and the references given there.

[317] The motion of these two diameters[1] about the two small[2] circles is not[3] about their centers, but rather about an eccenter[4], similar to the situation of the motion of the star[5] in longitude.[6] The returns[7] of these two diameters about the two small circles is according to the motion in longitude* . [318] Their motion[1] in the quarters[2] is also according to it[3], so that when the star moves through one of the quarters of its circle in longitude, and it returns to the nodal point[4] or one of the northern or southern[5] extremities[6], these two small diameters move through a quarter of the small circle.

[319] The motions which are particular[1] to the star Mercury are nine: (1)[2] The motion of its entire orb in the order of the signs around the center of the ecliptic orb and on its poles. It is one part in each hundred years.[3] This motion is called the motion of the apogee. (2) The motion of the eccentric dirigent orb[4] of the deferent[5] orb against the order of the signs.[6] It [makes] one revolution in each year.[7] It is called the motion of the dirigent [orb]. (3) The motion of[8] the deferent orb in the order of the signs on its own special[9] fixed poles. This motion is called the motion of the center of the epicyclic orb. (4) As this motion faces the ecliptic circle, it is the motion which is called the motion in longitude. (5) The motion of the epicyclic orb about its center and on its fixed poles which sets in motion the body of the star. It is called the motion in anomaly. (6) The motion of the dirigent line[10] of the diameter of the epicyclic orb about the fixed point [the equant]** . It is called the motion of equal distance. (7) The motion of the entire plane of the inclined orb to the plane of the ecliptic and its inclination from it.[11] It is called the motion of inclination. (8) The motion of the diameter of the epicyclic orb, that is, the one passing through the furthest distance and the closest distance[12] on the small circle, inclining to the

---

* The motions of the epicyclic diameters (see preceding note) must keep pace with the planet's motion in longitude, especially in order to insure the various coincidences at the quadrants that are described in the next passage. Therefore, the motions of these diameters are not about the centers of the small circles, but about eccenters (still, of course, located in the correct plane) whose eccentricity is equal to the planet's eccentricity; and the rate of these motions is equal to the rate of the planet's motion in longitude.

** See our note to passage 328.

north and to the south. It is called the inclination of the epicyclic orb. (9) The motion of the diameter of the epicyclic orb which intersects this diameter about the small circle, also to the north and to the south.[13] It is called the slant of the epicyclic orb.[14]

[320] Let us imagine a line drawn from the center of the world[1] to the center of the star, reaching rectilinearly to the surface of[2] the highest orb.[3] Let us also imagine a great[4] circle drawn on the poles of the ecliptic orb, passing through the endpoint of the line. This circle thus intersects the circumference of the ecliptic circle.[5] The point at which it intersects it is called the position of the star on the ecliptic circle. The arc of this circle between the endpoint of the line and the point of intersection is that which is called the latitude of the star. [321][1] This summary is the accurate presentation of the situation with regard to the orbs of Mercury. This is the diagram for the circles which we have mentioned.

[273] 1. round: - (L1).

2. parallel: - (H3).

3. surface: - (H6, 8).

4. the orb of: - (H2).

[274] 1. This passage is missing in L1.

2. the orb of: - (H1, 5, 7).

[275] 1. that is, the orb of Mercury: - (L1).

2. from west to east: *ab oriente in occidentem* (L1).

[276] 1. round: - (L1).

2. sphere: - (H8).

3. and set apart: - (H6, 8: L1).

4. parallel spherical: - (H2).

5. is removed from: H6, 8 read *ha-yoṣe* for *yoṣe* . The addition of the definite article changes the phrase to read: which is the center of the sphere eccentric to the world.

6. from the plane of: from the center of (H2 - 5, 7).

7. orb: + the sphere (H6, 8).

8. upper: - (H2 - 5, 7).

9. common: - (all translations)

[277] 1. round: - (H2 - 5, 7: L1).

2. is separated from this orb: - (H3, 7).

3. orb: + *ha-sovev li-galgal ha-nose* (which turns the deferent orb) (H4, 5).

4. parallel: - (L1).

5. which is. . . sphere: - (L1).

6. is removed: *ha-yoṣe* (the eccentric) (H6, 8). Cf. note 5 to passage 276.

7. the center of the dirigent orb: *merkaz ha-sovev* (the dirigent center) (H1 - 5, 7) = *ha-galgal ha-m*ᵉ*sovev* (the dirigent orb). (H6, 8).

8. Its distance. . . is: - (H2).

9. of the world: - (H4, 5).

10. surface: + *istius spere* (L1).

11. of the dirigent orb: - (H2 - 4, 7).

12. This eccentric orb is called the deferent orb: There are numerous and important variants to this phrase whch must be noted. The wording of my English translation is founded upon the version found in Arabic K; I have chosen it because it emphasizes the fact that Mercury has several eccentric orbs. Arabic Y reads: This orb is called the eccentric-deferent orb. L1 chooses a more economical wording: *Et dicitur orbis deferens* . The Hebrew manuscripts all follow our text (Arabic K), though there is a minor scribal error in H1 (*mi-yoṣe* for *ha-yoṣe*).

13. other than: - (L1).

14. with a circular motion: - (H6, 8).

[278] 1. parallel: - (L1).

2. fixed: - (H3, 4; L1).

3. exterior: H6, 8 read *ha-nir'eh* (visible), which is another meaning of the Arabic *ẓâhir* .

4. parallel: *egredientes* (L1).

5. common: - (all translations).

6. diameter: + of this orb (H6, 8).

7. parallel: - (H3 - 5, 7).

8. in two. . . surfaces: - (H2).

9. sphere: - (H2 - 5, 7).

[279] 1. the body of: - (L1).

2. the star, that is: - (L1).

3. the surface of : - (L1).

4. epicyclic: *nogah* (Venus) (!) (H3, 7).

5. Its epicyclic . . . orb: - (H1).

[280] 1. special: + fixed (H2 - 5, 7).

2. entire: - (H4, 5; L1, 2).

4. and it carries. . . center: - (H6, 8; L1); about its center: of the Hebrew manuscripts, only H1 reads like the Arabic. H4, 5

change the phrase to read "with the motion of the center"; H7 reads "it carries its center"; H3 has "it carries the center of its center" (*sic*).

5. imaginary: - (L1).

6. from the motion of its center: from its motion (H3 - 5, 7).

7. It: *hic circulus* (L1).

[281] 1. parallel: - (H2 - 5, 7).

2. deferent: eccentric (H2).

3. which are. . . orb: - (L1); The Hebrew manuscripts add: It also produces in the surface of the dirigent orb two circles parallel to the first circle. Their center is the center of the dirigent orb. H6 reads "in the surfaces" for in "in the surface." Moreover, H2 - 4, 6-8 are missing "to the first circle": H6, 8 are missing the remarks following "to the first circle" as well.

[282] 1. Two. . . first orb: - (H6, 8).

2. Two . . . of the world: *et similiter in superficie orbis primis* (L1); whose center is the center of the world: - (H4, 5).

3. upper: the firt (H6, 8).

4. which. . . surface: - (L1).

5. This circle: - (L1).

6. the inclined orb: - (H6, 8).

[283] 1. A circle. . . circle: - (H6, 8).

2. This circle. . . orb: - (L1).

[284] 1. deferent: eccentric (H2).

2. of the extremity of the line: - (H2 - 5, 7).

3. the dirigent orb: *ha-galgal ha-merkazim ha-meni ᶜa* (literally: the orb of the centers which sets in motion) (H6, 8).

4. furthest distance: *aux* (L1).

5. the orb of Mercury: *orbis excentrici* (L1); of the dirigent orb (H2).

6. the world: the earth (H2).

7. closest distance: *hadid* (L1).

[285] 1. which encompasses. . . orbs: - (L1).

2. circular: - (H6, 8).

3. the orbs of this star: - (L1); this star: *kotav* (H1), *kokhav* (H2), both of which are Hebrew names of Mercury.

4. them: - (H1).

5. the furthest distance: - (H7).

6. the furthest distance and the closest distance: *et augem et hadid* (L1).

7. and the two. . . intersection: - (H6, 8).

[286] 1. This motion is a slow motion: - (H6).

2. in the order. . . Its period: - (L1).

3. mentioned. . . *Almagest* : - (L1); mathematics: + known as the *Almagest* (H6, 8). Note that in the Arabic, the *Almagest* is called Ptolemy's "book on mathematics:

4. circle: *ha- ᶜolam* (the world) (H6, 8).

[287] 1. of Mercury's: - (L1) ; of the stars' (H3).

2. fixed: - (L1).

3. deferent: eccentric (H2).

4. and all that is in it: - (L1).

[288] 1. special: - (L1).

[289] 1. the circumference of: (L1).

[290] 1. This passage is missing from L1. This and the following passage are conflated in H4, 5.

2. whole: - (H7; L2).

[291] 1. motion: + by means of its motion (H6, 8).

2. the body of the star, i.e., : - (L1).

3. with a circular motion: - (L1).

[292] 1. imaginary: - (L1).

2. its center: the center of the star (H6, 8).

3. whose center. . . orb: - (L1).

4. the epicyclic. . . epicyclic orb: - (H2).

5. is inclined: - (H4, 5); *'asher* (that) (H6)

6. is inclined. . . orb: - (H3).

[293] 1. of the star: + the epicyclic orb (H1).

2. always: - (H2, 5, 7).

3. and when. . . signs: - (H5).

[294] 1. When. . . its diameter: *Diameter vero deferentis* (L1); diameter: + rectilinear (H1).

[295] 1. Let us. . . and: - (H4, 5).

2. furthest distance: *auge* (L1).

3. epicyclic orb: the entire (H1).

4. The circle. . . orb: - (L1).

[296] 1. which is. . . distance: - (L1).

2. deferent: *cicunferentem* (!) (L1); - (H2).

[297] 1. it sets. . . orb: - (H1).

2. it circles around: *we-yani ᶜehu* (it sets it in motion) (H6, 8).

3. center: - (H6, 8).

4. where it was at first: *ad primum punctum* (L1).

[298] 1. as we said earlier. However: - (L1).

2. moves: - (H6, 8).

3. against: *circa* (L1).

4. solar: - (L1).

5. which turns it: - (L1).

6. Its diameter. . . coincide: H1 rephrases this somehwat awkwardly to read: The pole (sic; an error of one letter, writing *qoṭev* for *qoṭer*) which passes through the center of the deferent orb returns to that which turns it from full coincidence with the first line.

7. the furthest distance and the closest distance: *aux et hadid* (L1).

8. also. . . description: - (L1); H6, 8 read here: until it returns to coincidence in the first place.

[299] 1. which turns: - (H1, L1).

2. in the order of: against the [order of] (H4, 5).

3. moves. . . signs: - (L1)

4. this period of time, i.e., : - (L1).

5. will coincide: *yasbiv* (will circle around?) (H6, 8).

6. the center of the deferent orb: - (H1).

7. in this year: - (L1); in the solar year (H6, 8).

[300] 1. the furthest distance: *auge* (L1).

2. the closest distance: *augis oppositam* (L1).

3. similar. . . moon: - (L1).

[301] 1. epicyclic orb: *galgal ha-ha ᶜataqah* (orb of motion) (H1) (!). Note this unusual term. Words built on the Hebrew root *ᶜ.t.q.* are used to translated forms of the Arabic *n.q.l.* and have the connotation of local motion.

2. solar: *tᵉmimah* (complete) (H6, 8).

[302] 1. It is also said: - (H1).

2. always: - (L1).

[204] 1. motion: *tᵉmunat* (the shape of) (H2).

2. the motion. . . orb: - (H3). This phrase is missing from Arabic Y and is found in the margin of Arabic K.

3. also keeps. . . upper and: - (H4, 5).

4. only: - (H6, 8).

5. only from: *nisi ab auge epicycli et ab oppositio augis, scilicet* (L1).

6. very: - (H6, 7).

7. the midpoint: + *ha- nᵉqudah* (the point of) (H5).

8. which is. . . line: - (H6).

9. always: - (L1).

10. For this. . . sun: - (H6; L1).

[305] 1. circle: orb (H1 - 5, 7).

2. in the order of: against the [order of] (H1, 2).

3. However, given this: *tunc* (L1).

4. its epicyclic orb: + the center of (H3 - 5, 7).

5. in the order of: *'el hefekh* (H3) = *bi-hefekh 'orekh* (H6, 8). Both mean: against the [order of].

[306] 1. which is called: - (H2, 5, 7).

2. against: *circa* (L1).

3. whole: - (H6, 7; L1)

4. Its motion: + *planete* (L1); motion: + other (H6, 8).

5. epicyclic orb: + with the deferent orb (H6, 8).

6. accidental: - (H4) ; *b^e-hannaḥah* (by assumption or by placement) (H3 - 5, 7).

7. it: the star (H6, 8).

8. measure: + for example (H6, 8).

9. As long. . . description: - (L1).

10. retrograde: - (H1); *'eḥad* (one) (H6, 8).

[307] 1. This passage appears in L1 as follows: *Et cum erit superius movetitur secundum ordinem et similiter epiciclus, et tunc videtur directus* .

2. the two: all (H8).

3. the two. . . signs: - (H2 - 5, 7).

[308] 1. the whole plane moves: *superficies declivit* (L1).

2. and it continues. . . ecliptic: - (H2).

3. Then it separates from it: - (H4, 5).

[309] 1. the furthest distance and the closest distance: *aus et eius oppositum* (L1).

2. of the epicyclic orb: - (L1).

3. is not always: H7 breaks off here rather abruptly, resuming again in passage 318; the manuscript was apparently misbound.

4. it moves: + *semper* (L1).

5. the circumference of: - (L1).

6. at its extremity: Arabic K adds here "nearer" (*al-aqrab*). L1 reads: *qui est apud oppositum augis* . H6, 8 read like Arabic K. H2- 5 read: near its two extremities, the furthest distance and the closest distance. H1 combines the readings of both H6, 8 and H2- 5.

7. The motion of: - (H2 - 5).

8. inclination of the epicyclic orb: H1 inserts in the middle of this phrase the word *mayl* , which is a transliteration of the Arabic term for "inclination."

[310] 1. as if it were: *qui est* (L1) = *w e-hi* (H1 - 5, 7). H2 adds here: *saviv* (around).

2. circle: orb (H6, 8).

3. speed: - (H4, 5: L1, 2)

4. the passage. . . orb: - (L1).

[311] 1. eccentric: - (H1).

2. furthest distance: *auge* (L1).

3. the inclined orb. . . intersection: - (H6, 7).

[312] 1. closest distance: *oppositam augi* (L1).

2. the furthest distance: *augem* (L1).

3. which. . . direction: - (H2 - 5); H1- 5 add: This inclination is to the north.

4. closest distance: *oppositus augis* (L1). H6, 8 add here *mi-galgal ha-nose min ha-merkaz* . Either the scribe intended *mi-galgal ha-yoṣe min ha-merkaz* (from the eccentric orb), or else *mi-galgal ha-nose* (from the deferent orb).

5. closest distance: *seu oppositio* (L1).

6. is complete: *erit maxima* (L1).

[313] 1. the epicyclic. . . intersection: - (H4, 5).

2. the inclined. . . orb: - (H6, 7).

[314] 1. When. . . intersection: - (H6, 7).

2. furthest distance: *augis* (L1).

3. always: - (H6, 8).

[315] 1. furthest distance: *ange* (L1). A typographical or scribal error for *auge* .

2. the plane of: - (H2, 4).

3. It drives: *we-yaḥzir we-yashuv we-yesovev* (H1). All three of these words can be considered more or less synonymous. *We-yaḥzir* is the translation of Ibn Paṭer for *qâda* ("to drive"), and *we-yesovev* is Ibn Tibbon's term for the same. *We-yashuv* , literally "to bring back" (and thus equivalent to *we-yaḥzir*), may also mean in this context "to do again"; it appears to be the addition of H1.

4. It then leaves it: H6 at this point runs, right in the middle of the sentence, into the beginning of the next chapter, picking up there also in mid-sentence! (Thus the rest of this chapter is missing in H6).

5. Then it returns. . . always: *deinde redit etiam ad superficiem excentri. Post ea tendit versus meridiem, deinde redit et sic* . (L1).

[316] 1. when the: from the plane of (H8).

2. the head or the node: - (L1). My translation combines the readings of both Arabic manuscripts. Arabic Y reads: "the point of intersection, the head"; and Arabic K, "the point of intersection, i.e. the node."

3. on the circumference. . . circle: *super illum circulum predictum* (L1).

[317] 1. of these two diameters: *huius orbis* (L1).

2. small: - (L1).

3. not: - (H2).

4. an eccenter: *ha-yoṣe* (H8). The addition of the definite article implies that the motion is about the (i.e., the planet's) eccenter.

5. the star: the stars (H8).

6. but rather. . . longitude: H2- 4 read, "the motion of its center is like the motion of the star in longitude," but it is not clear which center is meant; in longitude: *sed secundum longitudinem* (L1).

[318] 1. in longitude: H8 adds here: as we mentioned concerning the returns of these diameters.

2. quarters: + in longitude (Arabic K); H7 resumes here.

3. according to it: *derekh mashal* (for example) (H8).

4. the nodal point: *ro'sh ha-tanin* (H1) = *ha-ro'sh* (H2 - 5, 7). All mean the "head of the dragon," i.e., the point where the planet crosses to the north in its motion in latitude.

5. northern and southern: - (H2 - 7). This phrase is misplaced in H1.

6. Their motion. . . extremities: *et similiter in quartis prima quod quando planeta movebitur per unam quartem, verbi gratia a capite draconis usque ad unam duarum sinuum* (?) (L1).

[319] 1. which are particular: - (H2, 5, 7; L1).

2. (1): The motions are numbered in Arabic Y (not without some problems), in L1, and H8.

3. years: + *secundum Ptolomeum* (L1).

4. eccentric dirigent orb: first eccentric, which is the dirigent (H8).

5. deferent: - (H8).

6. signs: H4, 5 add: on its own fixed poles.

7. It. . . year: - (H4, 5).

8. The motion of: - (L1).

9. special: - (L1).

10. line: *galgal* (orb) (H2).

11. and its inclination from it: - (L1).

12. that is. . . closest distance: - (all translations).

13. It is called. . . south: - (L1).

14. It is called. . . orb: *we-tiqore galgal n ᵉṭiyyat ha-haqqafah* (it is called the orb of the slant of the epicycle) (H1). H3-5, 7 are missing *n ᵉṭiyyat* (here: slant). This ninth motion is placed by H2 after motion no. 7.

[320] 1. from the center of the world: - (H2).

2. the surface of: - (H2, 3, 7).

3. Let us. . . orb: (H4, 5).

4. great: - (all H except H8).

5. This circle. . . ecliptic circle: - (L1).

[321] 1. L1 reads for this passage: *Ista est dispositio orbium mercurii* (L1).

Chapter Twelve

The Orb of Venus

[322] The orb of Venus is a round[1], spherical body, surrounded by two parallel surfaces whose center is the center of the world. The highest of the two is tangent to the orb of the sun. The lower surrounds[2] the orb of Mercury. The parecliptic orb is in its upper surface.[3] This orb[4] moves slowly on the poles of the ecliptic orb in the order of the signs, similar to the motion of[5] the orb of[6] Mercury.[7]

[323] Another round[1] sphere is separated from this sphere. It is bounded by two parallel[2] surfaces[3] whose center, which the center of the sphere[4], is removed[5] from the center of the world. Its upper surface is tangent to the upper surface of the first orb in one point. It moves in the order of the signs with an equal[6] motion on poles fixed for it. This orb is called the deferent orb.

[324] A round sphere is separated from this sphere between its parallel[1] surfaces, which is the epicyclic orb.[2] It moves around its center on its fixed poles. The body or the star[3], i.e., Venus[4], is embedded in the body of this orb.

[325] When the deferent orb moves, an imaginary circle is produced[1] from its motion at the point of the center of the epicyclic[2] orb.[3] Concerning it we imagine the same as we imagined concerning the orbs of Mercury. Thus circles are produced in the orb of Venus similar to those which were produced in the orbs of Mercury, that is, the circles[4] parallel to the deferent orb and the inclined orbs.[5]

[326] Let us imagine a straight line drawn from the center of the world to the center of[1] the deferent orb, reaching to its circumference in both directions.[2] Its extremities are the furthest distance and the closest distance.[3] [327] Let us imagine also the diameter of the epicyclic orb to be on the rectilinear [extension] of this line, at the furthest distance.[1] Thus, when the deferent orb moves, the diameter of the epicyclic orb moves.[2] It faces a point other than the center of[3] the deferent orb[4], whose distance from the

center of the deferent orb is the same as the distance of this center from the center of the world.

[328] If we imagine a straight line running between this point and the center of the epicyclic orb on the rectilinear [extension] of that diameter, [then] this line is called the dirigent line of the diameter of the epicyclic orb. The point is called[1] the dirigent center and it is called the center of equal velocity* .

[329] The first orb moves on poles which face the poles[1] of the ecliptic orb. It carries the orbs of Venus, the plane[2] of the inclined orb and its furthest distance[3], and all that is in it[4], with a slow motion in the order of the signs, which is one part in each hundred[5] years.[6]

[330] The deferent orb moves on its own proper, fixed[1] poles, about its center with a circular motion[2] in the order of the signs. It sets in motion by means of its own motion the epicyclic orb. The body of the star also moves with the motion of the epicyclic orb.[3]

[331] The center of the deferent orb is fixed and does not move, except for the first, slow motion, contrary to the situation of the star[1] Mercury.[2] The epicyclic orb also moves on its poles about its center.[3] It sets in motion the body of the star, i.e.[4], Venus , about its center.[5]

[332] A circle[1] is produced from the motion of the center of the body of[2] the star whose center is the center of the epicyclic orb.[3] It is also called the epicyclic orb. The circle is inclined to[4] the plane of the eccentric orb[5], similar to the situation of the orb of Mercury. The motion of the center of[6] Venus is always on its circumference. When it is in its upper[7] [part], its motion is in the order of the signs. When it is in its lower[8] [part]. its motion is against the order of the signs.[9]

[333] The diameter of[1] the epicyclic orb whose endpoints are the furthest distance and the closest distance[2] moves on a small

---

* This point is more commonly known as the "equant"; the full Arabic term, *markaz mu ᶜaddil al-masīr* , may be rendered literally "the center which makes the passage[s] equal." For an account of the discovery of the equant see Neugebauer, *HAMA* , vol., pp. 154-156. Note that it appears that the concept originated from Ptolemy's analysis of the motion of Venus; thus the fact that the full technical term first appears in our text in the chapter on Venus (in the previous chapter, passage 319, the equant is called simply the "fixed point") offers further evidence just how closely Ibn al-Haytham is following the *Almagest* .

circle, driving with it the plane of the epicyclic orb[3], and, similarly, the diameter intersecting it at right angles, as in the figure of the orb of[4] Mercury.[5] However, the separation from the plane of the eccentric orb is at the closest distance[6] of the eccentric orb[7], [whereas] there [Mercury] it is from the furthest distance.[8] The separation of the intersecting diameter is there [Mercury] from one of the points of intersection, and here[9] [Venus] it is from the point opposite[10] it, which is the tail.[11]

[334] The motion of these two diameters about the two small circles is not about their center but about an eccenter[1], similar to the situation of the motion in longitude. The returns[2] of the two diameters are according to the returns[2] in longitude.

[335] The motion of the inclined orb is like[1] its motion for the star[2] Mercury. However, it is inclined to the north when the center of[3] the epicyclic orb is at the furthest distance and at the closest distance[4] of the eccentric orb. The inclination of the center of[5] the epicyclic orb[6] is always to the north, in contrast to the orb of[7] Mercury.

[336][1] There are eight motions which are proper to[2] the star Venus[3]: (1) the motion of its entire orb in the order of the signs with a slow motion[4]; (2) the motion of the deferent[5] orb in the order of the signs; (3) the motion which comes about from[6] this motion relative to the ecliptic circle, which is the motion in longitude; (4) the motion of the epicyclic orb about its [own] center[7]; (5) the motion of the dirigent line of the diameter of the epicyclic orb; (6) the motion of the inclined diameter of the epicyclic orb[8]; (7) the motion of the slanted[9] diameter of the epicyclic[10] orb, which intersects it[11]; (8) the motion of the entire plane[12] of the inclined orb.

[337] This summary is the situation of the orbs of Venus. However[1], we have abbreviated the chapter on the circumstances of this star because all of its motions and the form of its orbs are similar for the most part[2] to the orbs of Mercury[3], and we have just elaborated on their description in the discussion on the orbs of Mercury. All that which we have not presented[4] [fully] in this summary is, however, epitomized in it[5], given that it is set forth in the preceding [discussion]. This is the figure of the orbs of Venus.[6]

[322] 1. round: - (L1)

2. surrounds: - (L1).

3. The parecliptic. . . surface: - (L1).

4. This orb: + *totius* (L1).

5. the motion of : - (L1).

6. the orb of: - (H2, 3, 7, 8).

7. This orb. . . Mercury: In H4 this sentence reads: This orb moves on the poles of the ecliptic orb with (*b*^{e} -) the motion of Mercury.

[323] 1. round: - (L1).

2. parallel: + *kaduriyyim* (spherical)(H8).

3. It is bounded. . . surfaces: - (L1).

4. which is the center of the sphere: - (L1).

5. is removed: H3 has *ha-yoṣe* for *yoṣe* . The addition of the definite article implies that the center spoken of here is the eccenter of the planet.

6. equal: *sibbuvit* (circular) (H2 - 5, 7). The equivalent Arabic term, *mustadîrah* , is crossed out in Arabic K.

[324] 1. parallel: - (L1).

2. which is the epicyclic orb: - (H5).

3. star, i.e., Venus: All the Hebrew manuscripts have minor variants. The only one which changes the meaning is that found in H4, 5, which reads: of the stars, i.e., inclined Venus (!) (*noṭeh nogah*). This is probably due to some scribal confusion.

4. i.e., Venus: - (L1).

[325] 1. is produced: moves (H7).

2. epicyclic: - (H4, 5).

3. from its motion. . . orb: *ex motu centri epicicli* (L1).

4. circles: + mutually parallel (H8).

5. and the inclined orbs: There are many variants in the Hebrew tradition: H3, 7, read, "and they are the inclined orbs"; H4, 5 display a similar reading, changing only "orbs" to "circles"; and H2 has "and they are the deferent orbs."

[326] 1. the center of: - (L1).

2. in both directions: - (L1).

3. the furthest distance and the closest distance: *aux et eius oppositum* (L1).

[327] 1. furthest distance: *auge* (L1).

2. [epicycle] moves: + *directe* (L1).

3. the center of: - (H2 - 4).

4. whose distance. . . deferent orb: - (H6). H8 adds here: to the center of the deferent orb.

[328] 1. of the epicyclic orb. The point is called: *w ^{e}-hu* (and it is) (H7).

[329] 1. poles: + fixed (H6, 8).

2. the plane: *in superficiem* (L1); + *kol* (the whole) (H6, 8).

3. furthest distance: *augem* (L1).

4. and all that is in it: - (L1).

5. hundred: - (H1).

6. which is. . . years: - (L1).

[330] 1. fixed: - (L1).

2. with a circular motion: - (L1)

3. The body. . . orb: *et corpus planete* (L1); - (H2); + the center of [the epicyclic orb] (H3- 5, 7).

[331] 1. the star: - (H2- 8).

2. The center. . . Mercury: - (L1)

3. about its center: - (L1).

4. the body. . . i.e.: - (L1).

5. It sets. . . center: - (H2).

[332] 1. circle: + small (H6, 8).

2. the body of: - (all translations).

3. whose center. . . orb: - (L1).

4. is inclined to: *ᶜoveret ᶜal* (H1). The Hebrew literally means "passes over" or, perhaps, "moves over."

5. eccentric orb: *ha-nose ha-merkaz* (!) (H3). *Ha-nose* is the term for the deferent, *ha-yoṣe ha-merkaz* the term for the eccentric.

6. The motion of the center of: -(H3); the center of: -(H2-5,7).

7. upper: + of its epicyclic orb (all H).

8. lower: H2- 5, 7 read *shafel rum* , the technical term for nadir, rather than just *shafel* , the Hebrew word meaning "lower"; + *galgal haqqafato* (its epicyclic orb) (H1 - 3, 5 - 7).

9. The motion . . . signs: - (L1); its motion. . . signs: - (H4, 5).

[333] 1. The diameter of: - (H4, 5).

2. the furthest distance and the closest distance: *aux et eius oppositum* (L1).

3. driving. . . epicyclic orb: - (L1).

4. the orb of: - (H6, 8; L1).

5. as in the figure of: *et sic diximus in* (L1).

6. closest distance: *oppositum augis* (L1).

7. of the eccentric orb: - (H1, 3 - 5, 7).

8. there [Mercury] it is from the furthest distance: of the Hebrew manuscripts, only 3 and 7 follow the Arabic exactly. H1 is missing the word "there"; H4, 5 read, "and there it is at

this [furthest?] distance"; H6, 8 add "there, i.e., Mercury, it takes place when it is at the furthest distance"; the separation . . . furthest distance: - (H3).

10. opposite: *ha-m* ^{e}*ḥattekhet* (intersecting) (H3, 7).

11. which is the tail: - (all translations). These manuscripts thus all follow Arabic K, which also omits these words.

[334] 1. an eccenter: H3 again adds the definite article.

2. returns: *revolutiones* (L1).

[335] 1. is like: - (H4, 5).

2. the star: + the inclined orb of (H6, 8).

3. the center of: -(all H).

4. furthest distance and the closest distance: *augem et oppositum* (L1).

5. the center of: - (H1 - 5, 7).

6. at the furthest diatance. . . orb: - (H7); of the center. . . epicyclic orb: - (L1).

7. the orb of: - (all H except H8).

[336] 1. The eight motions are correctly numbered in H6, 8, as they are also in Arabic Y.

2. which are proper: - (L1).

3. to the star Venus: - (H1); to the star: - (H2 - 7).

4. with a slow motion: - (H3 - 5, 7; L1).

5. deferent: - (H3).

6. which comes about from: *ha-roḥav* (!)(H3 - 5, 7). This is the Hebrew term for latitude. Here it is certainly a scribal error, perhaps a confusion of the word *ha-howeh* found in H1.

7. about its center: - (L1).

8. These two motions are somewhat confused in most of the Hebrew translations. H6, 8 correctly translate the Arabic as it stands in our text. H3, 7 are nearly correct, except that in motion (6) they omit the word "epicyclic" so that the text reads, "The motion of the diameter of the inclined orb (*galgal ha-noṭeh*)." The word *ha-haqqafah* (epicycle) is missing: *ha-noṭeh* (inclined) should be modifying *qoṭer* (diameter). The other Hebrew manuscripts conflate the two motions into one. H1 reads: The motion of the inclined dirigent line of the diameter of the epicyclic orb. H2, 4, 5 are similar, but omit "inclined." Thus H1, 4, 5 have only seven motions. H2 adds another below, thus bringing its total up to eight. It appears that Ibn Tibbon, or one of the earliest copiests of his translation, had some trouble with the admittedly very complex Ptolemaic theory for the latitude of the inferior planets. The copiest of H2 did what he could to display eight motions, the number stated in the text. See below, n. 11.

9. slanted: H6,8 display *ha-maqom* , an obvious miscopying of *ᶜaqum* , the word employed by Ibn Paṭer to translated the Arabic *munḥarif* . The Ibn Tibbon translation uses *ha-noṭeh* (inclined); as in the previous chapter, it has no special technical term for the slant of the epicyclic diameter.

10. epicyclic: - (H3, 4, 7).

11. the motion. . . intersects it: H2 makes this into two motions, "The motion of the diameter of the epicyclic orb," and

the problematic *t ^e^nu ^c^at galgal ha-ḥittukhim ha-noṭeh*, which literally means, "The motion of the inclined orb of intersections." H3 - 5, 7 also exhibit *ha-ḥittukhim*.

12. entire plane: - (H3).

[337] 1. This. . . However: - (L1).

2. for the most part: - (H1, 4, 5).

3. and we have. . . Mercury: - (H2 - 5, 7).

4. which we have not presented: - (H1).

5. however, epitomized in it: - (H1).

6. on the circumstances. . .Venus: *et figure orbium assimilantur mercurii* (L1); This. . . Venus: - (H6, 8).

Chapter Thirteen

The Orbs of the Superior Stars

[338] The orbs of the three[1] superior stars, i.e., Saturn, Jupiter, and Mars, are alike in all their circumstances, in the number of their orbs, and the types of their motions, as Ptolemy explained in the *Almagest* * .[2] Therefore, because they are not very different [from each other], we have put the discussion of all of them into one chapter. [339] Each one[1] of these stars[2] has a sphere proper to it which is bounded by two parallel spherical surfaces whose center is the center of the world. Each sphere encompasses that which is adjacent to it.[3] [340] The first of them is the orb of Saturn. Its upper surface is tangent to the sphere[1] of the fixed stars[2], and its lower surface is tangent to the convexity of[3] the orb of[4] Jupiter. Likewise, the orb of[5] Jupiter is tangent to the orb of Saturn[6] with its upper surface, and its lower surface is tangent to the orb[7] of Mars. Similarly[8] for the orb of Mars: its upper surface is tangent to the orb of Jupiter, and its lower surface[9] is tangent to the orb of the sun.[10]

[341] Each one of these orbs moves slowly on poles which are alined with the poles of the ecliptic orb, as in the motions of[1] Venus and Mercury. [342] An eccentric sphere is separated from each one of these orbs.[1] It is surrounded by two spherical surfaces whose center is the center of the [eccentric] sphere.[2] It moves around its center and on its own fixed[3] poles with an equal[4] motion in the order of the signs. This sphere[5] is called the deferent orb.

[343] For each one of the three stars[1], a round sphere is separated from this[2] [deferent] orb between its parallel surfaces,

---

* Ibn al-Haytham must be referring here to Book X, chapter 6 (p. 480 in the new translation of G. T. Toomer, *Ptolemy's Almagest* (New York-Berlin-Heidelberg-Tokyo,1984)), where Ptolemy introduces his discussion of the superior planets with the remark that the model (but not the method for deriving the parameters) developed for Venus will hold good for all of the superior planets as well, and to Book XIII (Toomer, pp. 597-601), where, in treating of the latitude theory, the superior planets are taken together.

similar to what we presented earlier. Each one of them[3] is called the epicyclic orb of that star.[4] They move about their centers and on their poles with a circular motion. [344] The body of each of the three stars is embedded in the body of its epicyclic orb, and it moves with its [the epicycle's] motion.

[345] When the deferent orb moves with its proper motion, its sets in motion the epicyclic orb. Its [the epicycle's] center[1] describes an imaginary circle which is also called the deferent orb.[2] If we imagine its [circle's] plane[3] to intersect the world, the inclined circles and the circles[4] parallel to the deferent orb are produced, like the figure[5] in the orbs of Mercury and Venus. However, the inclined orbs for these orbs are fixed in one position. They neither move nor change[6] [place].

[346] Let us imagine a straight line drawn from the center of the world to the center of the deferent orb and extended rectilinearly to the circumference of[1] the inclined orb in both directions. It will then intersect the circumference of the deferent[2] orb in two points. One of them is the furthest distance and the other is the closest distance.[3] [347] The point of furthest distance[1] of each one of these lines[2] is always in the north, and the point of closest distance[3] is always in the south, because the inclination of the inclined orbs[4] of these stars neither changes nor moves. They are inclined[5] in one manner: the furthest distance to the north and the closest distance to the south.[6]

[348] Let us assume the diameter of the epicyclic orb to be on the rectilinear [extension] of this line. Then, as the deferent orb moves, carrying the epicyclic orb[1], its [epicycle's] diameter moves from[2] alinement with that line, and at that time it is found to be facing[3] a point other than the center of the deferent[4] orb and the center of the world, as in the case of the orb of Venus. [349] If a straight line is drawn between that point and the center of the epicyclic orb which [line] is on the rectilinear [extension] of the diameter[1] of the epicyclic orb, this straight[2] line is called the dirigent[3] line of the diameter of the epicyclic orb of the star[4], as in the case of the orbs of Venus and Mercury.

[350] The first orb of each one of the three[1] stars[2] moves on its own poles which are alined with the poles of the ecliptic orb with a slow motion, which is one part in each hundred years.[3] All that is

contained within it of the orbs of its star moves with its motion.[4] It carries the whole[5] plane of the inclined[6] orb, the furthest distance[7], and the center[8] of the deferent orb.

[351] The deferent[1] orb[2] moves about its center with an equal[3] motion that is in the order of the signs. It sets the epicyclic orb in motion by means of its own motion. The body of the star moves with the whole of the epicyclic orb.[4]

[352] The epicyclic orb moves around its center, setting in motion the body of the star.[1] An imaginary circle is produced from the motion of the center of the star whose center is the center of the epicyclic orb. It is also called the epicyclic orb.[2] This circle is inclined to the plane of the eccentric orb, as in the case of the two previous[3] stars.[4] [353][1] The center of each of these stars moves on the circumference of this circle. When it is in its upper [part], it [moves] in the order of the signs, and when it is in its lower [part], it is against the order of the signs.[2]

[354] The diameter of this circle whose endpoints are the furthest distance and the closest distance[1] of the epicyclic orb also moves on a small circle. It is sometimes inclined to the north and sometimes to the south. It drives with it the plane of the epicyclic orb.[2]

[355] The separation[1] of the diameter from the plane of[2] the eccentric orb takes place when the center of the epicyclic orb is in the northern point of intersection, i.e., the head.[3] When it separates, it inclines[4] first to the north. [356] The motions of these diameters about the small circles are not about their center, but rather about other, eccentric centers[1], as in the case of the motion in longitude. [357] The returns[1] of the diameters about the small circles are equal in time to the returns in the ecliptic circle.[2] Similarly, the period of their motions in the quarters[3] coincides[4] with the motion in longitude in the quarters of the ecliptic circle.

[358][1] The motions which pertain to each one of the superior stars are six: (1) the slow motion of the entire orb in the order of the signs; (2) the motion on its own poles of the deferent[2] orb, which turns the epicyclic orb in the order of the signs; (3) the motion which comes about from this motion relative to the ecliptic circle, which is the motion in longitude; (4) the motion of the epicyclic orb about its center[3]; (5) the motion of the dirigent line of the diameter of the

epicyclic orb about the dirigent center[4]; (6) the motion of the[5] inclined diameter of the[6] epicyclic orb.[7]

[359][1] This is the accurate presentation of the circumstance of the orbs of the superior stars. It is a general exposition[2] of them, like the exposition of the first stars. For that reason, we have preferred to be brief rather than lengthy.[3] This is the figure of their orbs. It appertains to each one of the three stars.

Chapter Title : Only H6, 8 follow the Arabic exactly (which, as usual, is found only in K). H1 has "The Superior Stars"; H2, "The Three Superior Stars and their Orbs"; H3, 7 have "The Orbs of the Three Superior Stars," and H4, 5 "The Orb of the Three Superior Stars."

[338] 1. three: - (L1).

2. as Ptolemy. . . *Almagest* : - (L1).

[339] 1. One: - (H2, 3, 7).

2. these stars: *istium trium* (L1).

3. Each sphere. . . adjacent to it: - (L1).

[340] 1. sphere: orb (H6, 8).

2. the fixed stars: *kadurim rabim qayyamim* (H7). Literally, "many fixed spheres." This reading is crossed out in H3, with the correct text in the margin.

3. the convexity of: - (all translations). Note that this word is missing in Arabic Y.

4. the fixed stars. . . convexity of the orb of: - (H4, 5); and its lower surface. . . of the orb of: - (H1).

5. the orb of: - (H1).

6. the orb of Saturn: - (H4, 7).

7. orb: *kadur* (sphere) (H1 - 3, 7).

8. the orb of Mars. Similarly: - (H4, 5).

9. is tangent to. . . lower surface: - (H4, 5).

10. Similarly. . . sun: (H6, 8).

[341] 1. the motions of: - (L1).

2. whose center. . . sphere: - (L1).

3. fixed: - (H3).

4. equal: *sibbuvit* (circular) (H6, 8).

5. sphere: - (L1).

[343] 1. In each one of the three stars: - (L1); stars: + *ᶜelyonim* (superior) (H2 - 5, 7). H1 has *ᶜinyanim* (issues, matters), which is probably a scribal error for the same word.

2. from this: *mi-merḥaq* (from the distance) (H3 - 5, 7). A puzzling error.

3. similar. . . Each one of them: - (L1).

4. of that star: - (L1).

[345] 1. Its center: + the point of (H1 - 5, 7); + of the epicyclic orb (H6, 8).

2. orb: *circulus* (L1).

3. plane: its two surfaces (H1 - 5, 7); + *istius circuli* (L1).

4. inclined circles and the circles: These terms are found in the singular in H8.

5. figure: - (H1) ; *tᵉnuᶜat* (motion) (H4, 5). This should be *tᵉmunat* ; *sicut diximus* (L1).

6. in one. . . change: - (L1).

[346] 1. the circumference of: - (L1).

2. the deferent: the inclined (H6, 8).

3. furthest. . . distance: in auge et in eius opposito (L1).

[347] 1. point of furthest distance: aux (L1).

2. lines: *ha-niqqudim* (!) (H4, 5). This is the word for "points", in an unheard of masculine plural form. It is not inconceivable that the text was being dictated to the copyist, who then misspelled *ha-niggudim* (the opposites), although the latter variant also is not found in the manuscript tradition.

3. point of closest distance: *oppositum eius* (L1).

4. inclined orbs: the inclined stars (!)(H1, 5); the three (H6, 8).

5. inclined: + always (H6, 8).

6. because. . . south: - (L1).

[348] 1. as the deferent. . . epicyclic orb: *cum ergo movebitur epiciclus per motum deferentis* (L1); carrying the epicyclic orb: - (H3, 4).

2. from: *ᶜal* (on) (H2 - 5, 7).

3. and at that time. . . facing: *ad verticandum* (L1).

4. deferent: the eccentric (H2).

[349] 1. the diameter of: - (H1).

2. straight: - (L1).

3. turning: *ha-mesev ha-sovev* (H1). Both terms are used to translate the Arabic *mudîr* . This sort of redundancy is not unusual in H1.

4. of the stars: - (H6, 8).

[350] 1. three: - (L1)

2. stars: - (all H except H1).

3. which is. . . years: - (L1).

4. of the orbs of its star: - (L1); *mi-galgalo* (of its orb) (H1- 5, 7). The word may also be read *mi-galgalaw* (of its orbs). H6, 8 read like the Arabic, missing only the possessive "its".

5. whole: - (L1).

6. inclined: deferent (H3).

7. furthest distance: *augem* (L1).

8. and the center: *b^{e}-merkaz* (on the center) (H6, 8) , i.e., about the center of the deferent.

[351] 1. deferent: eccentric (H1).

2. The deferent orb: - (H2).

3. equal: *ha-sovevet* (H6, 7). The word should be *ha-sibbuvit* (circular).

4. It sets. . . orb: H1 has a confused reading of this passage, which seems to mean: The epicyclic orb, with the whole epicycle [and?] the body of the star, moves with its motion; by means of. . . orb: - (L1).

[352] 1. The epicyclic orb. . . star: - (H2).

2. center of the star. . . center of the epicyclic orb: *centri principale* (L1); It is called the epicyclic orb: - (H1 - 5, 7).

3. previous: Venus and Mercury (H6, 8); fixed (H7).

4. as in the case. . . stars: - (L1).

[353] 1. This passage is missing from L1.

2. and when. . . signs: - (H4, 5).

[354] 1. the furthest distance and the closest distance: *aux et eius oppositum* (L1).

2. It drives. . . orb: - (L1).

[355] 1. The separation: + the beginning of (H1 - 5, 7).

2. the plane of: - (H7).

3. northern. . . head: H1 - 5, 7 jumble the word order here so that the text reads: The point of intersection, i.e., the northern head. We note that "head" is not found in Arabic Y.

4. When it separates it inclines: *movetur* (L1).

[356] 1. centers: *capita* (L1).

[357] 1. returns: *haqqafot* (H6, 8) = *revolutiones* (L1).

2. in the ecliptic circle: with slightly different wordings, H1 - 5, 7 all read "in the eccentric circles".

3. quarters: *partis* (L1).

4. coincides: *m esakkemet* (agrees) (H6, 8). This is not the same word which Ibn Paṭer usually uses to translate *inṭabaqa* .

[358] 1. The motions are correctly numbered in H6, 8.

2. deferent orb: + *ᶜim kol ha-ḥissaron ha-matḥil mi-mennah ᶜal husamah ben she-lifnei zeh* (H1). This addition means: with all the deficiency which begins [perhaps: arises] from it on its placement, whether before this. This obstruse phrase seems to be incomplete and is not found in any other Hebrew manuscript.

3. about its center: - (L1).

4. about the dirigent center: - (H2; L1).

5. the motion of the: - (H2).

6. diameter of the: - (H1).

7. diameter of the epicyclic orb: - (H2).

[359] 1. In place of this entire passage, L1 has simply: *Ista ergo dispositio planetarum superiorem breviter* .

2. exposition: H1 - 5, 7 display *hevdel* , a very literal translation of the Arabic *tafṣîl* . The full Arabic phrase is *jumlatu tafṣîlihâ* .

3. we have. . . lengthy: This sentence is translated in quite different fashion by Ibn Paṭer and Ibn Tibbon, though both convey accurately the meaning of the Arabic. Interestingly enough, H1 exhibits both translations (with minor variants) side by side.

Chapter Fourteen

The Orb of the Fixed Stars

[360] The orb of the fixed stars is one round[1] sphere surrounded by two spherical[2] surfaces[3] whose center, which is the center of the sphere[4], is the center of the world.[5] The higher of the two surfaces is tangent to the greatest orb which surrounds all the orbs and which moves with the swift motion.[6] The lower of the two surfaces is tangent to[7] the orb of Saturn. [361] This orb moves from west to east[1] in the order of the signs on fixed poles with a slow motion, which is one part of the parts of which 360 comprise the circle, in each hundred years.[2] Its poles are the poles of the ecliptic circle that the sun describes.[3] [All this is] according to what Ptolemy asserts to have found by his observations and the observations of those who preceded him* .[4]

[362] All the fixed stars are embedded in the body of[1] the orb[2] and do not leave their positions. Therefore, they do not change their positions[3] relative to each other. However, they do move as a whole[4], moving with the slow[5] motion of their orb[6] in the order of the signs.[7] [363] The center of each one of them produces a circle by means of the motion of this orb.[1] These circles are all parallel. Their poles are the poles of the ecliptic circle.[2] Each one of them moves always on the circumference of its circle[3] and does not leave it.[4]

[364] The motion of each one of the stars relative to the ecliptic circle and its position on it is[1] like the case of the rest of the stars.[2] That is, we imagine a circle drawn on the poles of the ecliptic orb, passing through the center of the star, and intersecting the world[3], the ecliptic circle which is on the surface of the highest orb, and also intersecting in a point the ecliptic circle which is on the surface of the ecliptic orb.[4] [365][1] That point is the position of the star on the ecliptic circle which is on the surface of the ecliptic orb.[2] The point

---

* The two salient points of Ptolemy's theory, namely, that the rate of precession is about one degree in each hundred years, and that this motion takes place about the poles of the ecliptic and not the poles of the equator, are discussed in *Almagest* , Book VII, chs. 2-3.

facing it on the ecliptic circle[3] which is on the highest orb is its true position in longitude. The motion of that point, which moves by means of the motion of the circle passing through the poles, is the motion of the star in longitude.[4]

[366][1] The line drawn from the center of the world[2] to the center of the star[3] which reaches to the surface of the highest orb is also in[4] the plane of that circle[5] and its endpoint is on the circumference of that circle.[6] That point is the position of the star on the surface of the highest orb. [367][1] The arc of the circle passing through the poles[2] which lies between this point and the position of the star on the ecliptic circle is the latitude[3] of the star.

[368] The whole orb[1] of the fixed stars is divided by six circles drawn on its poles. They divide it into twelve equal sections, which are the sections which we mentioned earlier, in the chapter on the whole orb. Each one of the sections lies between two semi-circles of these six circles.[2] [369][1] These circles also divide the ecliptic circle which is on the surface of this orb into twelve equal sections. The twelve sections of the whole orb are called signs. The arcs of the ecliptic circle which are in it are also called signs.[2]

[370] Each one of the sections of the whole orb[1] is called, as we said earlier[2], by the name of[3] the pattern [formed by] the stars which are in it. The arcs[4] of the ecliptic circle are called by the names of those signs. The arcs of the ecliptic circle which are on the surface of the highest orb[5] are also called by the names of those signs.[6] The sections of the highest orb are called by the names of these sections.[7] The sections of all the orbs of the stars[8] [i.e. the planets] are also called by the names of those sections* .

[371] In this orb there are a great number of fixed stars[1] of diverse size, fixed[2] in the entire body[3] of this orb.[4] Nevertheless, some of their positions have been observed, their place on the ecliptic circle has been learned, and their latitude is known.[5] That number is 1022, according to what Ptolemy said in the *Almagest* . He divided them into categories and collected[6] in each category those

---

* Ibn al-Haytham goes here to some trouble in order to explain how, for example, a planet can be said to have the longitude x degrees in Aries, even though the stars which form that constellation are in the highest orb whereas the planet is in its own distinct orb. This may also be an indication of the low technical level of Ibn al-Haytham's intended readership.

of them which are of equal magnitude.[7] [372] He said that 15 stars together make up the first magnitude, which is the greatest of them in magnitude. In the second magnitude there are 45 stars; in the third magnitude there are 208[1] stars; in the fourth magnitude there are 474[2] stars; in the fifth magnitude there are 217[3] stars; in the sixth magnitude there are 49[4] stars. There are nine dim and five nebulous stars**.[5] He mentioned the positions[6] of these on the ecliptic circle and their latitudes relative to the ecliptic circle. [373] He also compiled them in a manner such that one can determine their number and know their positions on the surface of[1] the orb.[2] That is, he envisaged the patterns that each group of these stars formed[3], and then he gave that constellation the name of that form. He then grouped the stars which he counted[4] into 48 forms and gave to each one of them the name of the animal or something else which it resembles. Thus all of the observed[5] fixed stars are comprised by the 48 forms and the stated number.[6]

[374] What [may exist] besides this [catalogue] is[1] [either] extremely small[2] [or else belongs to] that which is hidden from the people of the northern habitations due to its proximity to the south pole[3]; thus they have not been counted, and their position[4] on the ecliptic circle is not know. This is the figure of the orb of the fixed stars.[5]

** These subtotals are believed to be later interpolations into Ptolemy's text. See Toomer, *Almagest* , p. 399.

Chapter title: L1 reads *orbibus* for "Orb".

[360] 1. round: - (L1).

2. spherical: - (L1).

3. surfaces: - (H4, 5).

4. which is. . . sphere: - (L1).

5. world: *ha-* ^{c}*elyon* (the highest) (H1).

6. which surrounds. . . swift motion: - (L1).

7. of the two. . . tangent to: - (L1).

[361] 1. from west to east: *ab oriente in occidentem* (L1).

2. which is. . . years: - (L1).

3. describes: + by means of its motion (H8).

4. to have found. . . him: - (L1).

[362] 1. the body of: - (H2 - 5, 7).

2. orb: *spere* (L1).

3. positions: distance (H2 - 5, 7).

4. and do not. . . as a whole: - (L1).

5. slow: - (L1).

6. their orb: *galgal ha-kokhavim* (the orb of the stars) (H2).

7. in the order of the signs: - (L1).

[363] 1. The center. . . this orb: *et unaquoque eorum quando movetur designat circulum* (L1).

2. circle: *galgal* (orb) (H4, 5, 8).

3. of its circle: - (H3).

4. Their poles. . . leave it: - (L1).

[364] 1. and its position on it is: - (L1).

2. stars: + *in longitudine et latitudine* (L1).,

3. and intersecting the world: - (H2).

4. That is . . . longitude: - (L1); passing through. . .ecliptic orb: - (H4, 5); the ecliptic circle. . . ecliptic orb: - (H2, 3, 7).

[365] 1. This passage is missing from L1.

2. which is. . . ecliptic orb: - (H6, 8).

3. circle: H3, 6 add "on the surface of the ecliptic orb".

4. The motion. . . longitude: - (H6, 8).

[366] 1. This passage is missing from L1.

2. the world: the circle (H3).

3. to the center of the star: This whole phrase is missing in H7; H2- 5 lack "the center of"; and H1 reads "stars" for "star."

4. to the surface. . . also in: - (H6).

5. circle: orb (H2).

6. and its endpoint. . . that circle: - (H2- 8)

[367] 1. This passage is missing from L1.

2. the poles: the two stars (H8).

3. latitude: *merḥaq* (distance) (H2, 3, 7) : *maqom merḥaq* (place of distance)(!) (H4, 5).

[368] 1. orb: This word appears in the plural in H3. This manuscript also repeats a few lines within this passage.

2. which are. . . circles: - (L1).

[369] 1. For this passage L1 reads: *et similiter zodiacum et vocatur signa* .

2. the arcs. . . signs: - (H4, 5).

[370] 1. sections of the whole orb: of the Hebrew manuscripts, only H6, 8 (Ibn Paṭer) follow the Arabic exactly. H1 reads, "all the sections of the orbs," H2 - 5, "the section of all the orbs," and H7, "the sections of these orbs"; of the whole orb: - (L1).

2. as we said earlier: - (L1); as we said earlier: *[b e ?] khol ha-ma'amar* (!)([throughout ?] the entire treatise) (H6, 8).

3. the name of: - (H1, L1).

4. the arcs: the section (H2, 3, 7).

5. the highest orb: H4, 5 are missing from here to the end of this chapter.

6. signs: All the Hebrew manuscripts except H1 read *ha-ḥalaqim* (the sections).

7. The sections. . . sections: - (H2, 3, 7).

8. all the orbs of the stars: the orb of the stars (H1); of the stars: - (H6, 8).

9. The arcs. . . sections: - (L1).

[371] 1. stars: - (H6, 8).

2. fixed: - (H1 - 5,7).

3. body: - (H3).

4. fixed. . . orb: - (L1).

5. Nevertheless. . . known: *tantum quedam ex eis considerate fuerunt in loco in respectu zodiaci in longitudine in latitudine et in quantitate* (L1); H6, 8 follow the Arabic exactly; H1 follows the Arabic with minor changes in word order, H3 has, "their position has been observed, and their latitude is known", and H2, 7 are like H3 but instead of *merḥavam* (their latitude) display *shi ᶜuram [w ᵉ-] merḥavam* (their size [and] latitude).

6. and collected: - (H2, 3, 7).

7. He divided. . . magnitude: - (L1).

[373] 1. 208: The alphanumeric in H6 seems to read 48.

2. 474: 274 (H1, 2). The number appears alphanumerically in H2 and is written out in H1.

3. 217: The alphanumeric in H8 looks like 219.

4. 4 9: 249 (H1, 2). This is the only number for which there is a discrepancy in the Arabic, for K exhibits 48.

5. He said. . . nebulous stars: *in maiore quantitate XV, secunda XIV, in tertia. . . in quarte. . . in quinta. . . in sexta. . . et occulte sunt et nubose 5* (L1). Millás notes, "este y los otros tres espacios en blanco aparcen en este pasaje del manuscrito."

6. positions: *distanciosearum* (L1). An obvious typographical error for *distancias earum* .

[373] 1. the surface of: - (H2, 7).

2. orb: + the highest (H1 - 3, 7).

3. He also compiled. . . formed: - (L1).

4. which he counted: - (L1).

5. observed: *ha-y ᵉdu ᶜim* (known) (H6, 8). H1 adds here *ha- m ᵉṣuyyar* (imagined; depicted).

6. Thus all. . . number: - (L1).

[374] 1. What. . . is: *alie vero stelle* (L1).

2. extremely small: *b ᵉ-takhlit ha-n ᵉṭiyyah* (of extreme inclination) (H1 - 3. 7), intending, as it seems, too great a [southerly] declination. H6 , 8 exhibit *b ᵉ-takhlit ha-qaṭavut* . The second word clearly is a miscopy of *ha-q ᵉṭanut* (smallness).

3. and that. . . south pole: *aut sunt prope polum mundum* (L1); all the Hebrew manuscripts run this phrase together with the preceding, yielding, "and due to their proximity to the south pole, are hidden."

4. position: *distancie* (L1).

5. This is. . . stars: - (H1, L1).

Chapter Fifteen

The Highest Orb

[375] The highest orb, which encompasses all existing things, surrounds the sphere of the fixed stars and touches it.[1] It moves on its poles, which are the poles of the world, with the swift motion from east to west. All[2] the orbs of the stars move with its motion, as we made clear in the beginning of this treatise.[3] No stars are in it.

[376] All the inclined circles which pertain to the rest of the stars[1], the circles which pass through the poles of the ecliptic circle from which the latitudes of[2] the stars are found[3], all the great circles that were discussed at the beginning of this treatise, the ecliptic circle [relative] to which the motion of all the stars[4] is measured, the endpoints of the lines[5] which are drawn from the center of the world[6] and pass through the centers[7] of the stars--all of them are imagined [to be] in the surface of the highest orb which surrounds all the orbs.[8] [Relative] to it are measured all the situations of the stars in motion, aspects, [solar-lunar] conjunctions[9], and [planetary] conjunctions. [377] The rest of the motions and the situations[1] of the aspects are taken relative to the ecliptic circle which is in the surface of the highest[2] orb, by means of the positions[3] of the points which we determine to be the positions of the stars in longitude.[4]

[378] The true[1] distances between the stars [are measured] on the great circles which pass through the center of the world and the endpoints of the lines on the surface of the highest orb which are the position of the two stars.[2]

[379] The [planetary] conjunctions[1] are measured by means of the circle which passes through the poles of the ecliptic orb and the endpoint of the line which is the position of the star on the surface of the highest[2] orb. Given this situation[3], if it passes through the endpoint[4] of the line which is the position of the other star[5], the result of this situation is that the positions of the two of them on the ecliptic circle is one [and the same] point, for the circle which passes through their position is one circle. The conjunctions[6] of the sun and

the moon are like this, as we demonstrated in its place in the discussion of the orb of the moon.[7]

[380] All the motions which exist in all parts of the world, according to what we understand through the extent of our endeavor in[1] [studying] the views of the mathematicians and their recorded[2] observations[3], and according to what has been found by investigation based on[4] that which the inquiry of those who are involved in this science has attained, especially Ptolemy--for we have relied on his investigation[5] and observations[6] for[7] all the heavenly motions[8]--in line with what we said earlier in their exposition, are 47 motions:[9]

[381] The motion of the greatest encompassing[1] orb from east to west, which is the swift motion.

The motion of the orb of the fixed stars from west to east, which is the slow motion.[2]

Eighteen motions for the three[3] superior[4] stars[5]; each one of them has six motions.

Two motions for the sun.

Eight motions for Venus.

Nine motions for Mercury.

Six motions for the moon.

Two motions for that which is below the orb of the moon, which are the motions of the heavy and the light.[6]

From among the totality of motions of each one of the seven wandering[7] stars, there is one relative[8], non-essential* motion, which is the motion relative to the ecliptic circle.[9]

[382] Whereas our discussion of each one of the orbs has been particular [to that orb], and its figure has been drawn separately[1], still [as a result] the orb of each one of the stars[2] is understood and represented by itself[3] . All the orbs could scarcely be represented together with their intersecting circles their tangent orbs.[4] It is necessary that we draw a figure in which we will describe all the orbs of the stars together[5] in order, according to their configurations and according to what we explained separately about their affair[6], in

---

* The Arabic expression is *lâ dhâtiyah* . The essential motion, that is, that which belongs to the planet itself, is the motion of its eccentric orb, from which the motion relative to the ecliptic comes about (*ta ᶜriḍ u*) as a by-product, so to speak. See, for example, chapter 12, passage 336.

order to understand from their totality[7] the form of the world in all that it encompasses and that is contained in it.[8] [383] So that the orbs will not be numerous and the figure thus incomprehensible to the reader, each orb is limited to one of its positions, and it is sufficient for us to indicate that the rest of its positions [can be inferred from this one position] through its motion.[1] [384] In order that the task be easier[1], we assume all the inclined circles of all the stars to be in one plane[2], and their furthest[3] distances also to be on one line[4], and the centers of the eccentric orbs all to be on that line. This is more understandable as a depiction[5], and no harm is done to the principles of the figures. [385] We also make the eccentric dirigent center of the diameters of the epicyclic orbs of the superior stars and Venus as one, and we make it near the center of the world so as to be more understandable. For the purpose of the figure is only as a likeness. Likewise we make the center of the deferent orb of all the stars except Mercury as one. This is the figure of the whole.[1]

Chapter Title: There is no chapter title in either Arabic manuscript, though the spacing in the manuscripts would seem to indicate that a separate chapter was intended. L1 also seems to have viewed this as a separate chapter, for it begins: *sequitur de orbis supremo* . All of the Hebrew manuscripts leave room at the end of chapter 14 for the diagram and begin chapter 15 with oversized letters, except for H4,5, which run the two chapters together.

[375] 1. and touches it: - (H6).

2. All: - (H5, 6, 8).

3. as we. . . treatise: - (L1).

[376] 1. All. . . stars: - (L1).

2. the latitudes of: - (H4, 5).

3. All . . .found: - (H7).

4. all the stars: H2-5 read "the star."

5. the lines: the stars (H3, 4, 7).

6. the world: the circle (H2).

7. centers: *capita* (L1).

8. which surrounds all the orbs: - (L1); the orbs: the stars (H3, 4).

9. conjunctions and conjunctions: *coniunctionibus* (L1).

[377] 1. the situations: - (H6); the motions (H8).

2. highest: *illius* (L1).

3. the position: *b e-hannaḥat* (H1- 5, 7). This word would ordinarily be translated "by assuming" . It is a literal translation of the Arabic.

4. which we. . . longitude: *qui est locus stelle, in longitudine respectu ipsius* (L1); in longitude: - (H1).

[378] 1. true: - (L1).

2. and the endpoints. . . stars: - (L1).

[379] 1. The conjunctions: *ha-qawim* (the lines) (H6, 8); + *terminatur* (L1).

2. highest: - (H3-8).

3. situation: H6, 8 add "i.e., conjunction".

4. of the line. . . endpoint: - (H2).

5. on the surface. . . other star: - (L1).

6. conjunctions: H1 uses *qibbuṣim* , which seems most consistent; all the other Hebrew manuscripts display *ḥibburim* .

7. as we. . . moon: *sicut diximus superius* (L1).

[380] 1. what we. . . endeavor in: (H2, 5, 7).

2. recorded: - (H6, 8); *ha-howim* (extant or (unlikely) current) (H1 - 5, 7).

3. in the views. . . observations: *secundum opiniones et considerationes quadrivalium* (L1).

4. based on: or according to (H1).

5. investigation: - (L1).

6. observations: *ṭa ᶜanotaw* (claims or contentions) (H6); the corresponding reading in H8 is unclear.

7. for: *hinneh* (Hence [all the motions are. . .]) (H1).

8. all the heavenly motions: *in omnibus corporum celestium* (L1).

9. according. . . motions: - (L1).

[381] 1. greatest encompassing: highest, which encompasses all existing things (H6, 8).

2. which is. . . east: - (H2).

3. three: - (L1).

4. superior: - (H2).

5. stars: + wandering (H6, 8); H1 adds "which are Saturn, Jupiter, and Mars"; H2 has a similar addition, using an acronym for the three planets.

6. six. . . light: *et VI motus lune scilicet ponderosi et leves* (L1). L1 has obviously skipped a line of the text.

7. wandering: - (H2 - 5, 7).

8. relative: - (H1).

9. there is. . . circle: *est etiam unus motus essentialis et est motus qui est in respectu zodiaci* (L1). This is the opposite of the sense of the Arabic.

[382] 1. has been drawn separately: The translations all read "we have drawn". Arabic Y also displays a verb in the active pefect, first person plural; perhaps another Arabic manuscript read *khaṭṭaṭnâ* .

2. the orb. . . stars: H1 - 5, 7 read: one orb of the stars; H6, 8 read: each one of the orbs of the stars.

3. Whereas. . . itself: *et quia non intelligitur ex figuratione cuiuslibet stellarum per se nisi* (?) *spera cuiuslibet per se* (L1).

4. with. . . orbs: This phrase is confused in H2 - 5, 7, which have *ᶜiggulei galgalehem* , literally: the [intersecting] circles of their orbs. H3 adds to this, "and their orbs intersect."

5. It is necessary. . . together: - (H3); together: H2, 4, 7 insert here: since the circles of their orbs intersect and their orbs are tangent; H2 is missing "intersect."

6. according. . . particulars: - (L1).

7. form their totality: *ex ea* (L1).

8. in all that. . . in it: - (L1).

[383] 1. each orb. . . motions: *sufficiet nobis unus situs in qualibet orbe, et una dispositio suarum dispositionem* (L1).

[384] 1. In order. . . easier: - (L1).

2. one plane: *sua superficie* (L1).

3. furthest: - (L1; H1 - 5, 7). This follows the variant in Arabic K.

4. line: + *quoniam primo erit manifestus* (L1).

5. This is. . . depiction: - (L1).

[385] 1. This is. . . whole: - (H5, 7; L1).

Appendix

An appendix which we found in the hand of the *shaykh* , may God lengthen his life, at the end of the treatise. We have copied it as we found it.

Since the stars are natural bodies, it is necessary that each one of them by itself be allowed to move with one natural motion. Since the stars are [made] of a simple substance, their motions must be regular and uniform. The motion of each one of them does not vary.

Since that which moves by itself, moving from one place to another, vacates one place and fills another, we cannot avoid one of two possibilities. Either the place which it had occupied is empty, or another one of the surrounding bodies fills it [its place]. It is impossible that its place be empty.

If bodies successively occupy the places through which they pass, it necessarily follows from this that the body or bodies in which they move be pierced. Since the body of the heavens is of such a nature that it does not admit of being affected, it is refuted that any of the stars move in this manner.

There remains no manner for their motion other than [the following]: Each one of them has a round body which moves by itself with a circular motion. The body of the star is fixed somewhere in that body, and the star moves with its motion. Neither it nor the body which sets it in motion requires a place bigger than the place in which the two of them are.

Since each one of them has various motions, and it is not possible for it [the star] itself to have more than one regular motion, it necessarily follows from this that each one of them has movers by which they [each one of the stars] are driven and whose motions they follow, each one of them [the bodies] setting them [the stars] in motion with one kind of those motions. Because of this, it follows that each one of them [the stars], or some of them, has more than one body [all of] which move by themselves and do not require a place greater than the one which they have. The stars attached to them move with their motions.

Because of all this, it is necessary that the stars be surrounded by many spheres, one inside another. Each one of them has a motion proper to it on poles fixed for it, and a fixed axis on which it moves. It does not require more than its own place.

Final part: The premises on which the construction of the orbs of the stars and all the bodies which move around the world are based are four. The first of them is that the natural body does not by its nature move with more than one natural motion. The second is that the simple natural body does not move with a varying motion, that is, it always covers equal distances in equal times during its revolution. The third is that the body of the heavens does not admit of being acted upon. The fourth is that the vacuum does not exist.

LEXICON

We list here some of the more important scientific vocabulary employed in our text. The listing follows the order of the Arabic alphabet. In the left-hand column we present the Arabic roots (in parentheses) and below them the forms found in our text. In the right-hand column we display on successive lines (1) the English, (2) the Latin, and (3) the Hebrew translations. Unless otherwise indicated the Latin and Hebrew words are those found in all the manuscripts. If the Latin texts differ, we have used (L1) to signify for the text published by J. Millás Vallicrosa (see p. 41). Words that are used only in the Alfonsine translation (Ms. Oxford Canon. Misc. 45) are indicated by (L2). If we have found more than one expression in the Hebrew translations, we have identified them as follows: (T) Ibn Tibbon; (P) Ibn Paṭer; (H1) Vatican 399; and, in the cases where some of the Hebrew copyists altered the technical vocabulary, (H2), (H3), etc., according to the key which we presented in the introduction, pp.35-36.

('. *th. r.*)

ta'thîr	effect *operacio* (L1) *shinnuy* (T) ; *roshem, rishum* (P)

('. *kh. r.*)

al-muta'akhkhirûn	the later [astronomers] *posteriores* (L1) *ha-'aḥaronim*

('. *r. ḍ*)

'arḍ	earth *terra* *'ereṣ*

('.f.q.)

'ufuq — horizon
orizon
'ofeq

('.q.l.m.)

'iqlîm — climate
climata
'iqlim

('.w.j.)

'awj — apogee
aux
govah rum (T); *govah* (P)

('.w.n.)

fî 'ân waḥid — simultaneously
in uno instanti (L1); *simul* (L2)
b e-rega c 'eḥad

(b.ḥ.th.)

bâḥithan — by research
------
derekh ḥaqirah

(b.kh.r.)

bukhâr — vapor
vapor
'ed (T); *qiṭṭor* (P); *'awir* (H4,5)

(b.d.'.)

mabda' — beginning*
principium
hathalah

---

* In a philosophical context the correct translation would be "principle."

(*b.r.j.*)	
burj	[zodiacal] sign *signum* *mazal*
^{c}*alâ ('ilâ) tawâlî al-burûj*	in the order of the signs *secundum ordine signorum* (L1); *secundum cursum signorum* (L2) ^{c}*al meshekh ha-mazalot* (T); *bi-' 'orekh ha-mazalot* (P)
'ilâ khilâf tawâlî al-burûj	against the order of the signs *contra ordinem signorum* (L1); *ad oppositum signorum* (L2) *l*^{e}*-hefekh meshekh ha-mazalot* (T); *l*^{e}*-hefekh 'orekh ha-mazalot* (P)
(*b.r.d.*)	
baruda	to cool down, become cold *infridor* *qarar* (T); *hitqarer* (P)
(*b.ṣ.r.*)	
'abṣâr	eyes, vision *visus* (L1); *oculus, aspectum* (L2) *r*^{e}*ut*
mubṣar	visual object *corpus visi* (L1); *res qui videtur* (L2) *ha-guf ha-nir'eh* (P)*

* The clause which contains this term is missing from the Ibn Tibbon translation; cf. ch. 10, passage 255, n. 3.

(*b.*^{c}*.d.*)	
bu ^{c}*d*	distance *elongatio* (L1); *elongatius, distancia* (L2) *merḥaq* (T); *roḥaq* (P)
al-bu ^{c}*d al-'ab* ^{c}*ad*	furthest distance *longitudo longissima* (L1); *longitudo longior* (L2) *ha-merḥaq ha-raḥoq*
al-bu ^{c}*d al-'aqrab*	closest distance *longitudo propinquissima* (L1); *longitudo propinquior* (L2) *ha-merḥaq ha-qarov*
(*b.y.n.*)	
bayyana	to demonstrate, make clear, explain *declaro* (L1); *probator* (L2) *be'er*
(*th.b.t.*)	
thâbit	fixed *fixus* *qayyam* (T); *yoshev,* ^{c}*omed* (P)
(*th.q.l.*)	
thaqîl	heavy *ponderosa, gravis* *kaved*
(*j.r.m.*)	
jirm	body *corpus* *gerem* , ^{c}*eṣem* (T); *guf* (P)

(*j.z.'.*)

juz'	part *pars* *ḥeleq*

(***j.s.m.***)

jism	body *corpus* *geshem*, ^{c}*eṣem* (T); *guf* (P)

(*j.m.*^{c}.)

'ijtimâ ^{c}	conjunction *coniunctio* *qibbuṣ* (T); *ḥibbur* (P)

(*j.m.l.*)

jumlah (1)	general remark, summary --------- *k* ^{e}*lal*
jumlah (2)	group, constellation *aggregatio* *k* ^{e}*lal* (T); *qibbuṣ* (P)

(*j.n.s.*)

jins	sort *genus* *sug*

(*j.w.z.*)

majâz	passage* *intersitus* (L2) ^{c}*over* (T); *ma* ^{c}*avar* (P)

(*j.z.h.r.*)

al-jawzharân	[the two] nodes *extremitates draconis* (L1); *puncta draconis* (L2) *ha-tanin* (T); *ha-t* ^{e}*li* (P)

(*j.w.h.r.*)

jawhar	substance, intrinsic nature *substancia* (L1); *natura* (L2) ^{c}*aṣmut* (T); ^{c}*aṣmiyyut* (P)

(*j.y.b.*)

jayb	sine *sinus* *meitar m* ^{e}*ḥuṣṣah* (T); *ḥaṣi yeter, yeter* (P)**
jayb mustawin	[erect] sine *sinus rectus* (L1); *sinus equalis* (L2) *m* ^{e}*ḥuṣṣah yashar* (T); *yeter shaweh* (P); *meitar m* ^{e}*ḥuṣṣah* (H1,2)

---

* This term appears but once in our text, in the description of Mercury's motion in latitude (ch. 11, passage 310). That motion is the result of three separate components, each of which is called a motion (*ḥarakah*) and, as it seems to me, Ibn al-Haytham introduced *majâz* to denote the combined motion that is observed to the north and south of the ecliptic.

** For this and the following two entries, the variety of terminology in the Hebrew tradition reflects (often literally) a similar variety in the Arabic; the corresponding Arabic terms are given e.g. by W. Hartner, "Al-Battânî," *Dictionary of Scientific Biography* , vol. 1 (New York,1970) 507-16, p. 511. See also the following note.

jayb ma ^{c}kûs *	versine *sinus conversus* (L1); *sinus reflexus* (L2) *meitar nazor* (T); *yeter hafukh* (P)
(*ḥ.d.d.*)	
ḥadda	to form a boundary, demarcate *termino* (L1);*termino, diffino* (L2) *higbil*
(*ḥ.d.b.*)	
muḥaddab	convex *convexus* *givnoni*
(*ḥ.r.r.*)	
ḥarr	heat *calor* *ḥom*
(*ḥ.r.f.*)	
'inḥirâf	slant *reflectio* (L1); *devius* (L2) *n ^{e}ṭiyyah* (T); *^{c}iqqum* (P)
(*ḥ.r.k.*)	
ḥarraka	to set in motion *moveo* *heni ^{c}a*

---

* This is the reading of Arabic Y. Ms. K displays *mankûs* ; both define the function in terms of the *sahm* . (See chapter 7, passage 167, and our note there). All three terms - *jayb ma ^{c}kûs* , *jayb mankûs*, and *sahm*- are known in the Arabic literature ; cf. S. Tekeli, "Ḥabash al-Ḥâsib," *Dictionary of Scientific Biography* , vol. 5 (New York,1972) 612-620, p. 612.

taḥarraka	to move *moveo* *hitno ᶜe ᶜa*
ḥarakah	motion *motus* *t ᵉnu ᶜah*
ḥarakah baṭî'ah	slow motion *motus tardus* *t ᵉnu ᶜah mit'aḥeret* (T); *t ᵉnu ᶜah k ᵉvedah, t ᵉnu ᶜah m ᵉvoshsheshet* (P)
ḥarakat al-jawzaharayn	motion of the nodes *motus draconis* (L1); *motus capitis et caude draconis* (L2) *t ᵉnu ᶜat ha-tanin* (T); *t ᵉnu ᶜat ha-t ᵉli* (P)
ḥarakat al-muḥâdhâh	prosneusis *motus verticationis* (L1); *motus oppositi* (L2) *t ᵉnu ᶜat ha-nokhaḥut* (T); *t ᵉnu ᶜat ha-diyyuq* (P)
ḥarakat al-ikhtilâf	motion in anomaly *motus diversitatis* *t ᵉnu ᶜat ha-ḥilluf* (T); *t ᵉnu ᶜat ha-shinnuy* (P)
al-ḥarakah al-mukhtalifah	unequal motion *motus inequalis seu inuniformis* (L1); *motus diversus* (L2) *ha-t ᵉnu ᶜah ha-mitḥalefet* (T); *ha-t ᵉnu ᶜah ha-mishtannah* (P)
ḥarakah mustadîrah	circular motion *motus spericus* (L1);*motus rotundus* (L2) *t ᵉnu ᶜah sibbuvit*

al-ḥarakah al-dhâtiyyah	intrinsic or proper motion *motus essentialis* (L1); *motus proprius* (L2) *ha-t^{e}nu^{c}ah ha-^{c}aṣmit*
al-ḥarakah al-sarî^{c}ah	the swift motion *motus velocissimus* *ha-t^{e}nu^{c}ah ha-m^{e}hirah*
al-ḥarakah al-mustawiyyah	the equal motion *motus equalis* (L1); *motus medius* (L2) *ha-t^{e}nu^{c}ah ha-shawah* (T); *ha-t^{e}nu^{c}ah ha-mishtawah* (P)
ḥarakat al-ṭûl	motion in longitude *motus longitudinis* *t^{e}nu^{c}at ha-'orekh*
ḥarakat mu^{c}addil al-masîr	motion of equal velocity *motus equator cursum* (L1); *motus equatoris ambulatoris* (L2) *t^{e}nu^{c}at mashwah ha-mahalakh*
ḥarakat al-^{c}arḍ	motion in latitude *motus latitudinis* *t^{e}nu^{c}at ha-roḥav*
(*ḥ. ḍ. ḍ.*)	
ḥaḍîḍ	perigee *hadid* (L1); *hadid hoc est depressione* (L2) *shafel rum* (T); *shiflut* (P)
(*ḥ.w.r.*)	
miḥwar	axis *axis* (L1); *dyametrum* (L2) *qoṭer* (T); *^{c}amud* (P); *bari'aḥ* (H6)

(*ḥ.w.ṭ.*)

muḥîṭ (1)	surrounding, encompassing *continens* *maqif*
muḥîṭ (2)	circumference *circumferencia* (L1);*in circuitu*(L2) *maqif*

(*ḥ.w.l.*)

ḥâl	situation, circumstance *dispositio* *ᶜinyan*
al-istiḥâlât al-muta ᶜâqqibah	successive alterations *alteracio continuus* (L1) *ha-mishtannim ha-ba'im zeh 'aḥar zeh* (T); *ha-shinnuyim pa ᶜam 'aḥar pa ᶜam* (H6); *ha-shinnuyim ha-mitḥof ᵉfim ᶜalaw* (H8)*

(*kh.r.ṭ.*)

makhrûṭ	cone, conical *piramis* (L1); *pineata* (L2) *m ᵉḥuddad* (T);*maṣoq* (P); *midron* (H6)

(*kh.ṭ.ṭ.*)

khaṭṭ	line *linea* *qaw*

* I do not know which of the last two terms is that of Ibn Paṭer and which that of a copyist; the last expression does seem to conform to Ibn Paṭer's taste in Hebrew phraseology.

al-khaṭṭ al-mudîr
dirigent line
linea revolvens (L1);*linea motor* (L2)
ha-qaw ha-sovev (T); *ha-qaw ha-meni ᶜa* (P)

khaṭṭ al-istiwâ'
terrestrial equator
linea equalitatis (L1); *linea equinoctialis* (L2)
qaw ha-shiwwuy (T); *qaw ha-hashwayah* (P)

khaṭṭ niṣf al-nahâr
meridian line
linea meridiei
qaw ḥaṣi ha-yom

(***kh.f.f.***)

khafîf
light
levis
qal

(***kh.f.y.***)

khafî (1)
hidden
occultus
nistar (T); *ne ᵉelam* (P)

khafî (2)
dim
occultus (L1)
ḥashukh (T); *nistar* (P)

(***d.r.j.***)

darajah
degree
gradus
ma ᶜalah (T); *madregah* (P)

(*d.w.r.*)

dâ'irah	circle *circulus* ^{c}*iggul* , ^{c}*iggulah*
dâ'irat al-burûj	ecliptic circle *circulus signorum* (L1); *zodiacum* (L2) ^{c}*iggulat ha-mazalot*
dawâ'ir al-'azmân	time circles *circuli temporum* ^{c}*iggulei ha-z*^{e}*manim*
al-dâ'irah al-samtiyyah	zenith circle *circulus cenith et circulus elevationis, circulus cenitalium* (L1); *circulus zontium* (L2) *ha-* ^{c}*iggulah ha-n* ^{e}*khoḥit* (T); *ha-* ^{c}*iggulah ha-qodqodit* (P)
dâ'irat niṣf al-nahâr	meridian circle *circulus meridei* ^{c}*iggulat ḥaṣi ha-yom*
dawrah	revolution *revolucio* *sibbuv* (T); *haqqafah* (P)
mustadîr	round *rotundas* ^{c}*agol* (T); *m* ^{ec}*uggal* (P)
madâr al-jady	Circle of Capricorn* *circulus capricorni* *sibbuv g* ^{e}*di*

* In geography we call this the Tropic of Capricorn.

madâr al-saraṭân — Circle of Cancer
circulus capitis cancri (L1); *tropicus cancri* (L2)
sibbuv sarṭan

'istidârah — roundness
sperica (L1); *rotunditas* (L2)
^{c}*iggul*

(*r.b.*^{c}.)

tarbî ^{c} — quadrature
quadratura (L1); *quadratio* (L2)
ribu ^{c}*a* (T); *r* ^{e} *vi* ^{c}*iyyah* (P)

(*r.j.*^{c}.)

râji ^{c} — retrograde
retrogradus
ḥozer (T); *ḥozer 'aḥaronit* (P); *nazor* (H1)

(*r.kh.m.*)

suṭûḥ al-rukhâmât — sundials
superficies marmorum (L1)
sh ^{e}*ṭaḥim* (T); *sh* ^{e}*ṭaḥei avnei ha-shayish, sh* ^{e}*ṭaḥei ha-avanim* (P); *sh* ^{e}*ṭaḥei qaw ḥaṣi ha-yom* (H1)

(*r.ṣ.d.*)

'arṣâd — observations
rectifactiones (L1); *considerationes* (L2)
mabaṭim (T); *kiwwunim w* ^{e}*habbaṭot* (P)

(*r.f.*^{c}.)

'irtifâ ^{c} — altitude
elevacio (L1); *altitudo* (L2)
govah (T); *gavhut* (P)

(*r.k.z.*)	
markaz	center *centrum* *merkaz*
markaz mu ᶜ*addil al-masîr*	equant *centrum equans* (L1); *centrum equationis motus* (L2) *merkaz mashweh ha-mahalakh*
markûz	embedded, fixed *intromissus* (L1);*locatus, affixus* (L2) *taqu* ᶜ*a* (T); *nishqa* ᶜ (P)
(*z.ḥ.l.*)	
zuḥal	Saturn *saturnus* *shabtay*
(*z.m.n.*)	
zamân	season; time *tempus* *z* ᵉ*man* ; *y* ᵉ*mei* (H4,5, only in the sense of "season")
(*z.h.r.*)	
zuharah	Venus *Venus* *nogah*
(*z.w.y.*)	
zâwiyah	angle *angulus* *zawit*

^{c}alâ zawâyâ qâ'imah	perpendicular *secundum angulos rectos* *^{c}al zawit niṣevet*
(s.ḥ.b.)	
saḥâbiy	nebulous [stars] *nubose* (L1); *minor luminis, nubilose* (L2) *^{c}ananim* (T); *m^{ec}unnanim* (P)
(s.ṭ.ḥ)	
saṭḥ	plane; surface *superficies* *sheṭaḥ*
(s.m.t.)	
sâmata	to face, to be aligned with *respicio, directe respicio, vertico* (L1); *dyamtraliter , in directo [sum]* (L2) *hitnagged, nokhaḥ*
samt al-ra's	zenith *punctum verticilis, punctum cenith, cenith capitis* (L1); *punctum sumitatis capitis, zenith capiti, directum capitis* (L2) *nokhaḥ ha-ro'sh* (T); *govah ha-qodqod* (P)
(s.m.k.)	
sumk	width *spissitudina* (L1); *grossicius* (L2) *^{c}ovi*

(*s.m.w.*)

al-'ajrâm al-sâmawiyyah	the heavenly bodies *corpora supercelestia* (L1); *corpora celestia* (L2) *ha-g^{e}ramim ha-shama'iyim* (T); *ha-gufot ha-shama'iyim* (P)

(*sh.b.h.*)

'akthar shabahan	"more similar" [greater than]* *magis simile* (L1); *maior* (L2) *rav bi-dimyon* (T); *yoter* (P); *bi-dimyon gadol* (H8)
aqall shabahan	"less similar" [less than]* *minus simile* *m^{ec}aṭ bi-dimyon* (T); *qaṭan bi-dimyon* (P); *paḥot* (H8)

(*sh.r.y.*)

al-mushtariy	Jupiter *iouis* *ṣedeq*

(*sh.^{c}.^{c}*)

shu^{c}â^{c}	rays *radii* *niṣṣuṣ*

(*ṣ.m.t.*)

muṣmat	solid *plenum* (L2) *moqshi* (T); *miqsheh* (P)

* See ch. 4, passage 86 , where this term is defined.

(*ṣ.w.r.*)	
ṣûrah	form; diagram *figura* *ṣurah*
(*ḍ.d.d.*)	
mutaḍâdd	opposite *contrarius* *mitnagged, mithappekh*
(*ḍ.r.s.*)	
taḍârîs	corrugations, irregularities *denticitas* (L1) *ma ᶜaqashim* (T); *niqrot ha-ṣurim* (H6); *ḥafirot w ᵉ-n ᵉqiqim* (H8)*
(*ḍ.w.'.*)	
ḍaw'	light *lumen* *'or*
(*ṭ.b.q.*)	
inṭabaqa	to be in agreement with, coincide *applico, superpono* (L1); *locor* (L2) *hitdabbeq bi-shiwwuy* (T); *hitrakkev* (P)
(*ṭ.r.f.*)	
ṭaraf	end, extremity *extremitas* *qaṣeh* (T); *ro'sh* (P)

* I am unable at present to determine which of the last two terms is due to Ibn Paṭer.

(*ṭ.r.q.*)	
al-ṭarîqah al-muḥtariqah	"the burning path"* *via combusta* (L1);*zona torrida*(L2) *ha-r ^{e}ṣu ^{c}ah ha-nisrefet* (T); *ha-d ^{e}rakhim ha-nisrafim* (P)
(*ṭ.l.^{c}.*)	
ṭala ^{c}a	to rise *orientor* ; *ascendo* (L1) *^{c} alah*
al-ṭâli ^{c}	the ascendant *ascendens* *ha- ^{c}aliyyah* (T)** ;*ha-ṣome'aḥ* (P)
al-maṭâli ^{c}	the ascension [rising time] *ascensio* (L1);*elevationus* (!) (L2) *miṣ ^{c}ad* , *^{c}aliyyot* (T); *^{c}aliyyot* (P)
(*ṭ.w.l.*)	
ṭûl	longitude *longitudo* *'orekh*
(*ẓ.l.l.*)	
ẓill	shadow *umbra* *ṣel*

---

* See ch. 9, passage 208 and our note there.

** Clearly one would expect the form *^{c}oleh* .

(^{c}*.d.l.*)	
al-i ^{c}*tidâl al-rabî* ^{c}*iy*	vernal equinox *temperancie vernalis* (L1); *vernalis equationis* (L2) *ha-shiwwuy ha-avivi* (T); *n* ^{e}*qudat t* ^{e}*qufat nisan* (P)
al-i ^{c}*tidâl al-kharîfiy*	autumnal equinox *temperancie autumpnalis* (L1); *autumpnalis equationis* (L2) *ha-shiwwuy ha-ḥorpi* (T); *n* ^{e}*qudat t* ^{e}*qufat tishrei* (P)
(^{c}*.r.ḍ.*)	
^{c}*arḍ* (1)	accident *immutatio* (L1) *miqreh*
^{c}*arḍ* (2)	latitude *latitudo* *roḥav*
^{c}*araḍa*	to come about *provenior* (L1); *sum ab* (L2) *qarah* (T); *'ira* ^{c} (P)
(^{c}*.ṭ.r.d.*)	
^{c}*uṭârid*	Mercury *mercurius* *kokhav* (T) ; *kokhav ḥamah* (P); *kotav* (H1)
(^{c}*.ẓ.m.*)	
^{c}*uẓm*	magnitude *quantitas* (L1); *magnitudo* (L2) *godel*

(*ᶜ.k.s.*)

in ᶜikâs — reflection
conversio (L1)
hithapp ᵉkhut (T); *hippukh* (H6); *hitgarut* (H4,5)

(*ᶜ.l.m.*)

ᶜâlam — world (universe,cosmos)
mundus
ᶜolam

aṣḥâb al-ta ᶜâlîm — mathematicians
----
ba ᶜalei ha-limmudim (T); *ba ᶜale ha-ḥokhmot ha-limmudiyot* (H8)

(*ᶜ.m.d.*)

ᶜ amûd — perpendicular
perpendicular (L1);*dyameter* (L2
ᶜ amud

(*ᶜ.m.r.*)

al- ᶜimârah — the inhabited world
habitabilis (L1);*terra populata*(L2
ha-yishshuv

(*ᶜ.w.q.*)

ᶜ â'iq — obstacle
vetans (L1); *obstaculum* (L2)
me ᶜ'iq (T); *mone ᶜa* (P)

(*gh.r.b.*)

gharaba — to set
occido ; *descendo* (L2)
shaqa ᶜ

al-mughârib	descendant *occidens* (L1); *descendens* (12) *ha-shoqe* ᶜ*a*
(*f.ḥ.ṣ.*)	
faḥṣ	inquiry *inquisitio* *ḥaqirah*
(*f.r.ḍ.*)	
faraḍa	to assume *pono* *heni'aḥ*
(*f.r.q.*)	
fâraqa	to separate from, to leave *recedo* (L1); *separo* (L2) *hipared* ; *hitpared* (P)
(*f.ṣ.l.*)	
infaṣala min	to be separated from *seco* (L1); *separatur* (L2) *huvdal*(T);*hitḥalleq*(P);*huvrar*(H1)*
faṣl mushtarak	common section *directa communis* (?)(L1); *sectio communis* (L2) *hevdel m ᵉshuttaf* (T); *ḥittukh m ᵉshuttaf* (P)
(*f.l.k.*)	
falak	orb *orbis* (L1); *celum* (L2) *galgal*

* Note that the word chosen by H1 seems to imply a mental separation on the part of the inquirer.

falak al-burûj	ecliptic orb *orbis signorum* (L1); *signorum, zodiacum* (L2) *galgal ha-mazalot*
al-falak al-ḥâmil	deferent orb *orbis deferens* (L1); *celum deferens* (L2) *ha-galgal ha-nose*
al-falak al-khârij al-markaz	eccentric orb *orbis excentricus* (L1); *celum deferens* (*sic* in L2) *ha-galgal ha-yoṣe ha-merkaz*
falak al-tadwîr	epicycle, epicyclic orb *epiciclus* (L1); *epiciculus* (L2) *galgal haqqafah*
al-falak al-mudîr	dirigent orb *orbis revolvens vel equans* (L1); *celum motor* (L2) *ha-galgal ha-sovev* (*ha-mesev*)(T) *ha-galgal ha-meni*^{c}*a* , *ha-galgal ha-m*^{e}*sovev* (P)
al-falak al-a^{c}*ẓam*	the greatest orb *orbis maximus* (L1); *celum maior* (L2) *ha-galgal ha-gadol*
al-falak al-mumaththal bi-falak al-burûj	parecliptic orb *orbis similis orbi signorum* (L1); *celum simile zodiacum* (L2) *ha-galgal ha-domeh bi-galgal ha-mazalot* (T); *ha-galgal ha-nimshal bi-galgal ha-mazalot* (P)
al-falak al-mâ'il	inclined orb *declivis* (L1);*celum inclinatum*(L2) *ha-galgal ha-noṭeh*

(*q.b.b.*)	
al-qubbah	the dome* *torus* (L1); *collatorium* (?)(L2) *"al-qubbah"* ; *ṭabbur* (H8)
(*q.b.l.*)	
istiqbâl	opposition *oppositio* *niggud* (T); *mabaṭ* (P)*
(*q.r.'.*)	
istiqrâ'	investigation *inductio* (L1) *ᶜiyyun* (T); *ḥippus* (P)
(*q.r.n.*)	
iqtirân	conjunction (planetary) *coniunctio* *ḥibbur* (T); *qibbuṣ* (P)
(*q.s.m.*)	
qism, qismah	section, division *pars* *ḥeleq*
(*q.ṭ.b.*)	
quṭb	pole *polus* *qoṭev*

---

* See our note to chapter three, passage 61.

(*q.ṭ.r.*)	
quṭr	diameter *dyameter* *qoṭer*
(*q.^{c}.d.*)	
qâ ^{c}idah	base *basis* (L1);*basis fundamentus* (L2) *toshav*
(*q.^{c}.r.*)	
muqa ^{cc}ar	concave *concavus* *qa ^{c}aruri* (T);*^{c}aqum* (P);*kafuf*(H6)
(*q.l.b.*)	
al-inqilâb al-ṣayfiy	summer solstice *tropicus estivalis* (L1); *estavilis mutatio* (L2) *ha-hippukh ha-qayṣi* (T); *n ^{e}qudat t ^{e}qufat tammuz* (H6)
al-inqilâb al-shatawiy	winter solstice *tropicus hyemalis* (L1); *mutatio meridie, mutatio yemalis, inversio yemalis* (L2) *ha-hippukh ha-s ^{e}tawi* (T); *n ^{e}qudat t ^{e}qufat ṭevet* (H6)
(*q.m.r.*)	
qamar	moon *luna* *yare'aḥ* (T); *l ^{e}vanah* (P)

(q.w.d.)

qâda	to drive *redeo* (L1); *duco* (L2) *savav, hitqashsher* (!)(T); *ḥazar* (P)

(q.w.s.)

qaws	arc *arcus* *qeshet*
qaws al-samt	zenith arc *arcus cenith* *qeshet ha-nokhaḥ* (T); *qeshet ha-qodqod* (P)

(q.w.m.)

qâ'im	perpendicular *elevatus* ; *erectus* (L2) *niṣṣav* (T); *ᶜomed* (H6)
ᶜalâ al-istiqâmah	direct; rectilinearly *directe* *ᶜal yosher*
mustaqîm	straight *rectus* *yashar*

(q.y.s.)

ashkhâṣ al-maqâyîs	gnomons (shadow sticks) *gnomon* (L1); *quantitates sive mensure umbrarum* (L2) *maqish* (T); *qomah* (P)

(*k.th.f.*)	
kathîf (1)	heavy *grossius* (L1) *kaved* (T); ^{c}*aveh* (H8)
kathîf (2)	opaque *grossus* (L1); *spissus* (L2) ^{c}*akhur* (T); ^{c}*aveh* (P)
(*k.r.w.*)	
kurah	sphere *sphaera, spera* *kadur*
(*k.s.f.*)	
kusûf	eclipse *eclipsis* *laqut*
(*k.w.k.b.*)	
kawkab	star *stella, planeta* *kokhav*
al-kawâkib al-thâbitah	the fixed stars *stella fixa* *ha-kokhavim ha-qayyamim* (T); *ha-kokhavim ha-* ^{c}*omdim* , *ha-kokhavim ha-yoshvim* (P)
al-kawâkib al-sayârah	planets *erratica* (L1); *planeta* (L2) *ha-kokhavim ha-raṣim, ha-kokhavim ha-holkhim* (T); *ha-kokhavim ha-n* ^{e}*vokhim* (P)

(***k.w.n.***)

makân — place
locus
maqom

(***l.z.m.***)

lazima (1) — to keep to, adhere to
secundum sum, nunquam separatur a (L1); *maneo* (L2)
daveq li- (T); *shamar* (P)

lazima (2) — to necessarily follow [logic]
constat quod (L1)
hitḥayyev

(***m.r.kh.***)

al-mirrîkh — Mars
mars
ma'adim

(***m.s.s.***)

mumâss li- — tangent to
contingens
m ^{e}mashshesh (T); *noge ^{c}a* (P)

mutamâss — continguous
se contingo
mitmashshesh (L1); *noge ^{c}a* (P)

(***m.y.l.***)

mayl — declination, obliquity
declinatio ; *inclinatio* (L2)
n ^{e}ṭiyyah

(*n.ṣ.b.*)

muntaṣib	perpendicular *perpendicular* (L1) *niṣṣav* ᶜ*omed* (T); ᶜ*omed* (P)

(*n.ṣ.f.*)

niṣf	half *medietas* *ḥeṣi*
muntaṣaf	middle *medio* *ḥeṣi*

(*n.q.ṭ.*)

nuqṭah	point *punctum* *n*ᵉ*qudah*

(*n.q.l.*)

intaqala	to move, change position *moveo* (L1) *ne*ᶜ*etaq* (T); *hit*ᶜ*atteq* (P)
nuqlah, intiqâl, tanaqqul	motion, displacement *motus* *he*ᶜ*eteq*

(*n.h.r.*)

nahâr	day, daylight *dies* *yom*
niṣf al-nahâr	noon *meridie* *ḥaṣi ha-yom*

(*n.h.y.*)	
intahâ	to reach *pervenio* (L1); *applico* (L2) *kalah* (T); *higi* ^{c}*a* (P)
(*n.w.*^{c}.)	
naw ^{c}	type, sort, class *species* (L1); *modus* (L2) *min*
(*h.n.d.s.*)	
muhandisûn	geometers *geometres* *ba* ^{c}*alei ḥokhmat ha-shi* ^{c}*ur* (T); *ḥakhamei ha-shi* ^{c}*ur* (P); *ba* ^{c}*alei ha-m* ^{e}*handesim* (H1)
(*h.y.'.*)	
hay'ah	configuration ----- *t* ^{e}*khunah*
(*w.t.r.*)	
watar	chord *corde* *meitar*
(*w.j.h.*)	
jihah	region, direction *pars* *ṣad*

(*w.z.y.*)

muwâzin li-
parallel [homocentric]
equidistans (L1); *parallel* (L2)
n ^{e}khoḥi (T); *shaweh, mishtawweh* (P)

mutawâzin
mutually parallel [homocentric]
equidistans ad invicem (L1); *invicem parallel* (L2)
n ^{e}khoḥi (T); *mishtawweh* (P)

(*w.s.ṭ.*)

wasaṭ
middle, mean
medius
emṣa ^{c}

(*w.s.^{c}.*)

sa ^{c}at al-mashriq
rising amplitude
amplitudo orientalis (L1); *latitudo orientalis* (L2)
roḥav ha-mizraḥ

sa ^{c}at al-maghrib
setting amplitude
amplitudo occidentis (L1); *latitudo occidentis* (L2)
roḥav ha-ma ^{c}arav

(*w.ḍ.^{c}.*)

mauḍi ^{c}
place, position
locus
maqom

(*w.q.t.*)

waqt
instant
tempus (L1); *quando, hora* (L2)
^{c} et (T); *sha ^{c}ah* (P)

(*w.h.m.*)

tawahhama — to imagine
ymaginor
dimmah ; *ḥashav*, *ṣiyyer* (P)

INDICES

General Index

References to the introduction, apparatus, and lexicon (for those entries of the lexicon that have been included in the index) are given by page number and in regular type; references to the text of *On the Configuration* are given by passage number and in bold-face.

Arabic Word List

We list here those words that are found in the Introduction, and also some words that are cited in the apparatus to the translation but not included in the lexicon. However, we have not reproduced here all the words that are listed in the lexicon.

Hebrew Word List

We register here the noteworthy Hebrew words that appear in the introduction, the lexicon, and the the apparatus. Transcribing Hebrew is somewhat more complicated than Arabic; the full notation of philologists is cumbersome and unnecessary, yet one hesitates to sacrifice precision. We employ a somewhat simplified system of transcription that, we hope, is accurate but not pedantic; in particular: (1) long and short (and intermediate) vowels are not indicated ;(2) the *sh ^ewa* is signified by the letter "e" written above the line ; (3) we have followed current normal pronunciation in choosing which of the *b ^egadk ^efat* letters shall receive a separate transcription: thus we use both *b* and *v* for *beit* , but only *d* for *dalet* ; but otherwise each consonant has its unique notation, which is the same as that used in the transcriptions from Arabic; (4) we have not recorded the doubled consonants that result from the addition of the definite article. Our listing follows the alphabetical ordering employed in most Hebrew dictionaries, that is, verbs follow the root, but other forms beginning with *he* , *mem* , etc. are listed under those initial letters. Unusual words, usually found as a variant, are presented as they appear in our texts; only the more common verbs have been standardized to their infinitive forms. This prolegomenon notwithstanding, we are confident that the Hebraist shall without undue trouble be able to locate any item of interest.

Latin Word List

We have not tampered with peculiar spellings found in our texts, especially the occurrence of the letter *c* in place of *t* ; however, in most cases nouns and adjectives have been regularized to the nominative, and verbs to the first person singular.

ab [sum], 251
affixus, 246
aggregatio, 237
alteracio continuus, 242
altitudo, 245
amplitudo occidentis, 262
amplitudo orientalis, 262
anadir, 123
angulus, 246
applico, 249, 261
archus diei et noctis (!), 85
arcus, 78, 257
arcus cenith, 257
artificialis, 90
ascendens, 250
ascensio, 250
ascendo, 250
aspectum, 74, 235
astrologia, 125
autumpnalis equationis, 251
aux, 234
axis, 241

basis, 256
basis fundamentus (!), 256

calor, 239
celum, 253
celum deferens, 254
celum inclinatum, 254
celum maior, 254
celum motor, 254
celum simile zodiacum, 254
cenith capitis, 247

centri principale (!), 211
centrum, 246
centrum equans, 246
centrum equationis motus, 246
(in) circuito, 244
circuli principales ymaginari, 75
circuli temporum, 244
circulus, 244
circulus capitis cancri, 245
circulus capricorni, 244
circulus cenitalium, 244
circulus cenith et circulus elevationus, 244
circulus excentricus, 154,155
circulus meridie, 244
circulus mundi, 102
circulus signorum, 244
circulus zontium, 2444
circumferencia, 242
climata, 234
collacacio, 101
collatorium (?), 255
concavus, 256
coniunctio, 237, 255
coniunctio magna, 158
connexa, 74
considerationes, 245
consimilium, 66
constat quod, 259
continens, 242
contingens, 89, 259
se contingo, 259
contra ordinem signorum, 235
contrarius, 249
conversio, 252
convexus, 239
corde [chorda], 261
corpora celestia, 248
corpora supercelestia, 248
corpus, 236, 237
corpus visi, 235
cursum, 235

declaro, 236

Treatises and Manuscripts Cited

We have not listed again the manuscripts of *On the Configuration*

المقالة في هيئة العالم

لابي علي الحسن بن الحسن ابن الهيثم

****

حقّقها واجتهد في نشرها

الفقير الى رحمة ربّه

الدكتور إسحق صفي لنغرمان

****

شركة «غرلند»

نو يورك ولندن

١٩٩٠

الفهرست

تصدير المحقّق

باركي يا نفسي الله ، يا نفسي باركي الله! ولا تنسي شيئاً من آلائه، فإنّه الصافح جميع ذنوبك المشفي جميع أوجاعك الفاكّ من الفساد حياتك المتوّجك بالاحسان والرحمة. وقد تأخّر نشر هذا الكتاب مدة طويلة وقاسيت الأمرّين على اصداره تحمّلت عليه مشاق كثيرة ، فتحوا عليّ فا ُظلْمٍ وكالموني بلسان الباطل ، قد كافوني شرّاً بدل خير. ومع إنني عديم الشأن أصغر الباحثين وأطفّهم ومع ذلك فإنّ موضوع هذا الكتاب أعلى وأشرف ، وهو المقالة الهامّة «في هيئة العالم» لواحد من أكبر علماء الاسلام في القرون الوسطى ،وهو ابو علي الحسن بن الحسن ابن الهيثم، المشهور بمؤلّفاته الكثيرة والغزيرة في الرياضيات وخاصةً في علم المناظر وعلم الهيئة وعلم الهندسة. ومن أجل ذلك عندما قدّمت لي شركة «غرلند» - أدام الله العاملين فيها -اصداره انتهزتُ الفرصة وتقدّمتُ إليها بهذا الكتاب. وهو أطروحتي للدكتوراه مختصرة ومصحّحة بحسب الطاقة وبالله التوفيق.

ومع إنني لا استطيع أن اصف هاهنا مضمون المقالة «في هيئة العالم» كلّها كما أنني لا اتمكّن أن اعبّر عن خطورتها بنسبة تأريخ علم الهيئة في دار الاسلام وحتّى في سائر الأراضي أرى من الواجب أن اذكر في هذه الفاتحة العربية ولو بصورة مقتضبة النتيجة الرئيسية التي توصّلت إليها في بحثي ، بيد أنّ دلائلها موجودة في صدر الجزء الانكليزي. والمسألة المطروحة على بساط البحث هي : لأي غرض ألّف ابن الهيثم هذه المقالة؟ ومن المعلوم أن العلماء في القرون الوسطى قد ضاقوا بالمناقضة الظاهرة في صفة حركات الكواكب بين الفلاسفة التابعين لقول أرسطو وبين التعليميين (الرياضيين) التابعين لقول بطلميوس. يقول الفلاسفة إنّ حركات الأجرام السماوية بأسرها مدارات كراتٍ مجسّمةٍ متحرّكةٍ حركةً مستديرةً مستويةً حول مركز الأرض. ومن الناحية الأخرى يستعمل التعليميون دوائر ونقطاً متوهّمة بدل الأجسام المصمتة وبالاضافة إلى ذلك برهن بطلميوس في كتابه «المجسطي» أنّه من

الواجب أن نضع لهذه الحركات أدوات رياضية مثل «الخارج المركز» و«التدوير» ، التي ينفي الاستعمال بها قواعد الفلسفة الارسطوطالية. واليوم نعرف أنّه بمعونة هذه الأدوات نجح بطلميوس أن يقرب من القطع الناقص الذي هو الصورة الهندسية الموافقة لمدارات الكواكب السيّارة.

ونرى أن ابن الهيثم يقول في مقدّمته للمقالة «في هيئة العالم»: "نظرنا لكلّ واحدة من الحركات التي ذكرها [بطلميوس] في الوجه الذي به يحتمل أن تظهر تلك الحركة عن تحرّك جسم كري حركة بسيطة متّصلة دائمة"، كما أنّه يذكر في ضمن المقالة وخاصة في الجزء الثاني منها (فصول ٩-١٥) مراراً كثيرةً الاكر والأفلاك. ومنذ سبعين سنة تقريباً نظر المؤرّخ «دُوهم» في المقالة التي نبحثها هنّا - وحقّاً فإنّه لم ير إلاّ قطعا من الجزء الثاني، وفي الترجمة اللاتينية فقط - واستنتج أن غرض المقالة اصدار هيئة تامّة تشمل أجسام الفلاسفة ودوائر التعليميين ، تقليداً لما قام به بطلميوس في كتاب آخر له في علم الهيئة، «الاقطصاص». ومنذ نشر «دوهم» رايه لا يزال الباحثون يكرّرونه المرة تلو الأخرى.

وأنا متأكّد كلّ تأكيد أن «دوهم» أخطأ في تقديره ، وفي صدر الجزء الانكليزي قدّمتُ دلائل على أن مقالة ابن الهيثم تختلف كثيراً عن كتاب «الاقطصاص» من جهة النظم والمضمون والمصطلحات المستعملة فيه كما بيّنتُ أننا نجد في هذه المقالة كلّ المشاكل التي تقود إلى المناقضة المذكورة بين الفلاسفة والتعليميين، وهي نفس القضايا التي توجّه إليها ابن الهيثم في كتابه «الشكوك على بطلميوس». وفي رأيى إنّ المقالة «في هيئة العالم» ما هي إلاّ كتاب أساسي أوضح فيه ابن الهيثم جلّ الموضوعات المشتملة على علم الهيئة فيما عدا الرياضيات لأنّه خصّص هذه المقالة للمتعلّمين الذين لا يحذقون في الرياضيات . وقد سبقوه بتأليف هذا النوع من الكتاب في علم الهيئة، مثلاً كتاب الفرغاني، لكنهم لم يشرحوا الاتّصال والموافقة بين الدوائر المتوهّمة والأكر المجسّمة ، وفي المقالة «في هيئة العالم» يبيّن ابن الهيثم هذا المعنى أكمل مما بيّنه السابقون، لكنّه لم يحل المشاكل المرتبطة بهذا الأمر حتّى أنّه لا يذكر أهمّها ولو بكلمة واحدة. وليست مقالته محاكاة «الاقطصاص» على الرغم من زعم «دوهم» ، ولا نجد فيها هيئة جديدة

عارية عن المشاكل المذكورة، وعلى الأصح نلتقي بها بيانا صريحا غير رياضي لهيئة العالم وفقاً لآراء علماء العصر.

وأخيراً نصف طريقنا في تحقيق النصّ العربي للمقالة «في هيئة العالم»، حيث استخدمنا مخطوطين : أحدهما ، وهو الوحيد الكامل ، محفوظ في المكتبة الهندي «إنديا أوفس» في لندون برقم لوث ٨٣٤ أوراق ١١٦-١٠١ ورمزنا إليه بحرف «ي» ؛ والثاني موجود في مكتبة كستمونو تحت رقم ٢٢٩٨ أوراق ٤٣-١ وهو ينقص فصلين ، «القول على فلك الشمس» و«القول على فلك القمر» كما ينقص جملا متعدّدة من فصل ٢ ومن فصل ٨. ورمزنا إليه بحرف «ك». وبعد أن انهينا عمل التحقيق علمنا أنّه يوجد في المغرب مخطوط ثالث (رباط ، مالك ٨٦٩١) وقد طلبنا تصويره إلاّ أن طلبنا لم يُجب.

وقسّمنا النصّ إلى فقرات ورقّمناها أيضاً وتناولنا الترجمة الانكليزية بنفس التقسيم والترقيم لكي يسهل على القارئ المقارنة بين الأصل والترجمة. وأضفنا علامات الوقف والهمزات وبعض الشكل. وكلّ كلمة وجدنا لها نسخة مختلفة في أحد المخطوطين أو عدّلنا النسخة الموجودة في الاثنين أضفنا بعدها علامة الكوكب (*) ؛ وفي آخر النصّ يجد القارئ جهاز التحقيق مطابقة لترقيم الفقرات وفيها دليل لكلّ ما أدخلناه من تعديلات أو اقتراحات.

ولم يسبقنا أحد بنشر هذا النصّ ما عدا التعليق الموجود في مخطوط «ي» فقط وقد حقّقه ونشره الباحث م. شرام في كتابه المذكور في فهرست الأصدار (أنظر الجزء الانكليزي صفحة ٥٠) ، صفحة ٦٩. وفي جهاز التحقيق للتعليق رمزنا إلى تحقيقه برمز «شرام».

بسم الله الرحمن الرحيم
ألعزة لله*
قول للحسن* بن الحسن بن الهيثم
في هيئة العالم

[١] قال* لم يزل كثير من أصحاب التعليم ممن أنعم النظر في علم الهيئة وأدرك حقائق الصور الموجودة لجملة العالم ولجل أجزائه يجرّدون قولا مرسلا يشرحون* فيه بالجليل من القول كيفية شكل العالم وكل واحد من أجزائه الأول وأحوال جميع الأجرام السماوية وترتيب بعضها عند بعض وأبعاد بعضها من بعض ومقادير أجرامها واختلاف أوضاعها وأنواع حركاتها إلاّ أنّهم سلكوا في ذلك طريقا مطابقا في الظاهر للدقيق من النظر في علم الهيئة و منتزعا من تلك الأقاويل. [٢] ولأنّ تلك الأقاويل أعني الدالّة على صور الشكل وقوانين الحركات بالأرصاد المحققة والبراهين المصححة إنّما هي مبنية على حركات نقط متوهّمة على محيطات دوائر معقولة على ما يتبيّن في الموجود لنا من كتبهم ومفروضة كذلك المسلوكة في وجوده ليسهل لا غير عُكر مجسّمة هي المتحركة* بتلك* النقط تلك الحركات صار قولهم (٢١/ك) فيما شرحوه مقصورا على تلك الدوائر والنقط فقط [٣] إذ لم يجعلوا قصدهم ايضاح الطريق الذي به يمكن أن تتمّ* تلك الحركات المختلفة مع فرضها في سطوح أكر مجسّمة ولا كيفية حال الأكر الحاملة لها مع اختلاف مراكزها بل جعلوا أغراضهم التقاط ما ذكر منها بالتفصيل من الأحوال وايرادها جملا مجرّدة من البراهين والطرق المسلوكة في وجوده ليسهل لا غير على من هم أن يستفيد معرفة هيآت تلك الحركات متسلّما لا باحثا و مقلّدا لأهل* الصناعة لا* مستبصرا فهم* ما أوردوه من ذلك وتخيل ما حدوه ورسموه.

[٤] و ليست أقاويلهم مع هذه الحال في غاية الظهور في كل المواضع ولا مستوفية أيضا لجميع ما تقتضيه* المعاني بل مقنعة في بعض

المواضع متسمّحة في البعض. [٥] وليس قصدنا في جميع هذا القول طريق الطعن والتبكيت إذ كان ممكنا أن يكون أصحاب هذه الأقاويل وإن عدلوا عن المعاني التي أهملوها* متصوّرين لها قيمين بها لكن* متحرّين صفة الحال وممهدين بذلك طريقا للقول على هذه المعاني مع سبقهم إلى الكلام عليها ومبينين تغاير الأقوال فيها من بعض الجهات وإن (ب٢/ك) توافقت من بعض.

[٦] فنحن قائلون في هيئة العالم قولا يشتمل فيما نظن على كلّيات المعاني التي إليها انتهى ادراك الباحثين عن حقائق العلوم التعليمية ومعتمّدون بالقصد اقتصاص العلوم التي بيّنها بنهاية الاستقصاء بطلميوس الفاضل في كتابه في التعليم لكن نفرض للحركات الدوائر والنقط المتوهّمة التي أخذها مجرّدة في سطوح الأكر المتحرّكة بذواتها تلك الحركات إذ* كان ذلك أصدق من صفة الحال وأظهر بيانا* عند الأفهام. [٧] وكان غرضنا نقل ما فهمناه* من هذه العلوم إلى أفهام من يريد الوصول بغير بحث إلى ادراكها. [٨] وكان هو قاصدا فيما استعمله البراهين ومستغنيا فيها بتلك الدوائر والنقط فقط.

[٩] ولما كان قولنا بحسب ما بيّنه ورتّبه وكان معرضا عن استعمال شيء من الأجسام نظرنا لكل واحدة* من الحركات التي ذكرها في الوجه الذي به يحتمل أن تظهر* تلك الحركة عن تحرّك جسم كري* حركة بسيطة متّصلة دائمة. [١٠] ويمكن مع ذلك اجتماع جميع* تلك الأجسام المفروضة لكلّ واحدة من *(٢١/ك) الحركات معا من غير أن يعرض هناك تمانع ولا مدافعة ولا تعويق بل تكون* حركاتها مع اجتماعها دائمة متّصلة .[١١] فنفرض لكلّ حركة نذكرها بسيطة جسما كريا متحرّكا حول مركزه حركة متّصلة (ب١٠١/ي) ليكون ذلك أشبه بالأمر الدائم البعيد عن التغيير السليم من الآفات ثم نلخص القول في جميع ما هو موجود للأجرام السماوية وما يعرض لها من التنقّل* واختلاف الأوضاع.

[١٢] وقولنا في كلّ الحركات* إنّما هو بحسب رأي بطلميوس فيها واعتقاده وجميع ما نورده* إنّما هو جمل يفهم منها صورة كلّية

الشكل* والوضع والحركة لا جزؤيات هذه بل يحتمل متى قرن بها* جزؤياتها* أن تكون* مطابقة لها. واجتهادنا في ذلك بقدر الطاقة والامكان.

القول على جملة العالم

[١٢] العالم اسم موضوع يقال على جملة جميع الموجودات. وشكل العالم بكلّيته شكل كرة. والكرة شكل مجسّم يحيط به سطح واحد في وسطه نقطة كلّ الخطوط (ب ٣/ك) الخارجة منها إلى السطح المحيط به متساوية وتلك النقطة هي مركز الكرة*. [١٤] والعالم جسم مصمت شكله شكل كرة* يحيط به سطح واحد في وسطه نقطة كلّ الخطوط الخارجة منها إلى السطح المحيط متساوية وتلك النقطة* تسمّي مركز العالم. وأريد بقولي مصمت أنّه ليس فيه مكان خال* بل جميعه ملاء.

[١٥] والأجسام التي تملأ* العالم مختلفة الأنواع إلاّ أنّها تنقسم بالجملة إلى ثلاثة أقسام : إلى ثقيل وخفيف وما ليس بثقيل ولا خفيف. [١٦] فأما الثقيل فهو الذي يتحرّك من جميع نواحي العالم إلى المركز*. وأما الخفيف فهو* الذي يتحرّك من الوسط إلى جميع نواحي المحيط. وأما الذي ليس بثقيل ولا خفيف فهو الذي يتحرّك أبدأ حول المركز.

[١٧] وأما الجزء الثقيل من العالم فإن مكانه المختصّ به هو وسط العالم فلذلك صارت حركته من جميع نواحي* العالم إلى المركز ليستقرّ في مكانه الذي هو له بالطبع. وهذا الجزء الثقيل ينقسم قسمين هما الأرض والماء. [١٨] أما الأرض فإنّها بجميع أجزائها وما يتّصل بها من الحيوان والنبات ثقيلة بالطبع (٤١/ك) ومكانها الخاصّ بها هو وسط العالم الذي هو أقرب الأماكن من المركز. [١٩] وكلّ جزء منها إن رُفع* من مكانه* إلى مكان هو أبعد منه من المركز ثمّ خلي عنه فإنّه يتحرّك بما له من القوة الطبيعية إلى ناحية المركز ولا يزال كذلك إلاّ أنّ يعوقه عائق يحول بينه وبين المركز فحينئذ يقف حيث* لا يجد طريقاً إلى الحركة. [٢٠] فلو رُفع المانع الذي تحته لتحرّك إلى الموضوع الأخفض ولو أمكن فعل ذلك* دائماً لما زال متحرّكاً حتّى ينتهى إلى حيث يكون مركز العالم في وسطه ، وكذلك كلّ جزء من الماء.

[٢١] وشكل الأرض بكلّيتها وجميع أجزائها شبيه بشكل الكرة لكن سطحه ليس بصحيح الاستدارة بل فيه تضاريس* تعرض* فيه من تأثيرات حركات الأجرام* السماوية إلاّ أنّ ذلك ليس بمبطل لكريتها ولا بمخرج* عن شكلها بل هي بالاضافة إلى جملتها كالشخونة العارضة في سطح بعض الأكر الصغار. [٢٢] فالأرض بجملتها كرة مستديرة مركزها مركز العالم وهي مستقرّة في وسطه ثابتة فيه غير منتقلة إلى جهة من الجهات ولا متحرّكة بضرب* من ضروب الحركات [ب٤/ك] بل هي دائمة السكون.

[٢٣] فأما الماء فإنّه محيط بكرة الأرض إلاّ أنّه* لما كان ثقيلا وكانت* حركاته إلى المركز* وكانت كرة الأرض خارجة بينه* وبين المركز صار في أقرب الأماكن من المركز* فأحاط بكرة الأرض. [٢٤] ولما كانت الأرض متشعّثة* الظاهر وكانت فيها مواضع منخفضة ومواضع مرتفعة وكان* الماء من أجل* ثقله يطلب المركز وما قرب من المركز صار منحدرا بالطبع إلى المواضع المنخفضة فبقيت المواضع المرتفعة منكشفة كالجزائر التي تكون* في وسط البحر والماء يحيط* بها. [٢٥] وكلّ ما على الأرض من العمارات فهي في هذه الجزائر التي تكون* خارجة عن الماء*. [٢٦] وسطح الماء الظاهر سطح قطعة من* كرة صحيحة* الاستدارة متساوية* الأجزاء مركزه مركز العالم والخطوط الخارجة من مركز العالم إلى جميع النقط (١٠٢١/ي) التي على سطح كرة الماء متساوية وكذلك كلّ جزء من الماء وإن كان منفردا* كالغدرنات والمستنقعات. فهذا جملة القول على القسمين* الثقيلين.

[٢٧] فأما الجزء الخفيف من العالم فإنّ مكانه الخاصّ به هو فيما بين الجزء الثقيل (٥١/ك) والجزء الذي ليس بثقيل ولا خفيف*. [٢٨] ولذلك صار ما ينحصر منه في تضاعيف الجزء الثقيل يتحرّك صاعداً أبداً إذا ارتفعت العوائق إلى أن يلحق بهذا المكان أعني الذي فيما الجزئين الباقين ، وما هو ألطف منه* مترتّب على ما هو أكثف منه*. [٢٩] وهذا الجزء أعني الخفيف ينقسم* بقسمين* هما الهواء والنار. [٣٠] أما الهواء فهو محيط بكرة الماء وشكله شكل كري سطحه الداخل الذي يلي كرة الماء سطح مقعّر تقعيراً كرياً لانّ سطح الماء يحدّه إلاّ أنّ المواضع المرتفعة من الأرض تشعّثه كتشعّث* سطح كرة الأرض لكن ذلك غير مؤثر في كريته لصغر

قدر ذلك عند جملته. فأما سطحه الأعلى فسطح محدّب يلي كرة النار.

[٣١] فهذا* الجسم أعني الهواء مختلف الأجزاء يعرض فيه من عوارض* الكون والفساد وبعض أجزائه* كثيف وبعضها سخيف. [٣٢] أما الكثافة التي فيه* فللبخارات الغليظة التي تعرض* من سخونة الأجسام الرطبة وحركته صاعدة كالرياح العاصفة والسحاب المؤتلف* من البخار الرطب أو البرودة العارضة في* الأهوية الرطبة (ب٥/ك) وتجميدها أيّاها كالثلوج الحادثة وبالجملة فلأنواع الاستحالات المتعاقبة. [٣٣] والذي يلي الأرض من الهواء أكثر سخونة مما بعد عنها لانعكاس شعاعات* الشمس الواقعة على الجسم الكثيف ثمّ يزال الهواء يتزايد في البرد كلّما بعد عن الأرض حتّى يقرب من كرة النار ثمّ يعود ساخناً* للطافته وقربه من شبه النار وكلّما قرب من النار اشتدّ تشبّهه بها* إلى أن يصير إلى جرم النار لا تخالطه شيء من الهواء.

[٣٤] فأما النار فإنّها محيطة بكرة الهواء وشكلها شكل كري سطحها الداخل سطح مقعّر تقعيراً كرياً محيطاً* بكرة الهواء. فأما سطحها الخارج فسطح كري محدّب صحيح الاستدارة يحدّه* الجسم* المحيط به المتحرّك عليه. فهذا* جملة القول على الجسمين الخفيفين.

[٣٥] فأما الجزء الذي ليس بثقيل ولا خفيف فإنّه محيط بالجزئين الباقيين ومتحرّك حولهما حركةً دائمةً وهذا الجزء هو الذي يسمّى الفلك وهو الذي يشتمل على جميع الكواكب. [٣٦] وشكله بكلّيته* شكل كري يحيط* به ويحدّه (٦١/ك) سطحان كريان مركزهما مركز العالم أحدهما وهو الداخل سطح مقعّر صحيح الاستدارة مماس لكرة النار مركزه مركز العالم والآخر وهو الخارج سطح محدّب كري صحيح الاستدارة متاشبه* الأجزاء مركزه مركز العالم وهو محيط بجميع الموجودات. [٣٧] وهذا الجسم بجملته متحرّك حركةً سريعةً من الجهة التي تسمّى* المشرق إلى الجهة التي تسمّى* المغرب ويحرّك* بحركته جميع ما يشتمل عليه من الأجرام السماوية حركةً مستديرةً. [٣٨] و هذا الجسم ينقسم أقساماً وسنبيّن كلّ واحد من ذلك* في موضعه ، فهذا القول المجمل على جميع العالم وسنأتي على تفصيله إن شاء الله تعالى .

القول على الفلك

[٣٩] ألفلك اسم موضوع يقال على كلّ مقدار مستدير من جسم كريّ أو سطح كريّ أو سطح دائرة أو محيط دائرة فالجزء المحيط بالعالم المتحرّك حول المركز يسمّي بجملته فلكاً. [٤٠] وهذا الفلك ينقسم أقساماً كثيرةً إلاّ أنّه ينقسم إلى* تسعة أقسام هي أجسام كريّة متماسة يحيط أحدها* بالذي يليه ويماس السطح المقعّر من المحيط السطح المحدّب من المحاط به (٦١/ك) وهذه الأكر كلّها مركزها مركز العالم.

[٤١] وكلّ واحد من هذه يسمّي على انفراده فلكاً. [٤٢] فالأوّل منها فلك القمر ويليه فلك عطارد ويلي ذلك فلك الزهرة ثمّ يليه فلك الشمس ثمّ فلك المريخ ثمّ فلك المشترى ثمّ فلك زحل ثمّ فلك الكواكب الثابتة ثمّ الفلك المحيط* الأعظم المحرّك لجميع هذه الأفلاك*. [٤٣] ولهذه الأفلاك حركات مختلفة إلاّ أنّها (ب١٠٢/ي) تنقسم* بالجملة إلى حركتين متضادتين أحديهما من المشرق إلى المغرب والأخرى من المغرب إلى المشرق.

[٤٤] أما الفلك الأعظم المحيط بجميع الأفلاك فإنّه يتحرّك من المشرق إلى المغرب حركةً سريعةً على قطبين ثابتين ويدور في كلّ يوم وليله دورةً واحدةً ويتحرّك بحركته جميع ما يحيط به من الأفلاك. [٤٥] والقطبان الثابتان يقال لأحدهما القطب الشمالي لأنّه يلي الجهة التي تسمّي* الشمال وهذه الجهة هي الجهة العامرة من الأرض وهذا القطب هو القطب الظاهر عند أهل المساكن الشمالية. [٤٦] وأما القطب الآخر فيقال (٧١/ك) له القطب الجنوبي لأنّه يلي الجهة التي تسمّي* الجنوب وهذا القطب خفي عند أهل المساكن الشمالية.

[٤٧] وأما باقي الأفلاك فإن منها ما يتحرّك من المغرب إلى المشرق ومنها ما يتحرّك بخلاف ذلك أيضاً* ولكلّ* واحد منها* حركة بالجملة من المغرب إلى المشرق أو من المشرق إلى المغرب إلاّ أنّها على أقطاب مختلفة لكلّ واحد منها قطبان يخصّانها. [٤٨] وكلّ واحدة* من هذه الأكر تنقسم* بأكر

مختلفة لكلّ واحدة منها حركة تخصها* سنذكر كلّ واحدة منها* في موضوعها* إن شاء الله*.

[٤٩] وكلّ كرة تتحرّك* على قطبين فإنّ النقط التي عليها ترسم* دوائر في سطح الكرة قد بيّن ذلك المهندسون في كتبهم. وكلّ سطح يقطع أكراً متامستاً* فإنّه يرسم في كلّ واحدة* منها دائرةً قد تبيّن ذلك أيضاً. [٥٠] فإذ تحرّكت هذه الأكر على أقطابها رسم كلّ كوكب* فيها بنقطة مركزه دائرة متوهّمة في الكرة وإذا توهّمنا سطح تلك الدائرة قاطعاً للعالم فإنّه يحدث* في سطح الفلك الأعلى أيضاً دائرةً (ب٧/ك) وكذلك إذا توهّمنا سطحاً يمرّ بمركز الكوكب وبمركز* العالم ويقطع* العالم فإنّه أيضاً يحدث في سطح الفلك الأعلى* دائرة عظيمة مركزها مركز العالم. ففي سطح الفلك الأعلى دوائر كثيرة متوهّمة تحدث من هذين الجنسين لكلّ واحدة* منها صورة خاصّة*.

[٥١] فالفلك الأعظم إذا تحرّك الحركة السريعة فإنّ النقط التي عليه كلّها ترسم كما قدّمنا دوائر في سطح الفلك الأعظم وهذه الدوائر كلّها متوازية قطباها قطبا العالم وهي مختلفة المقادير وكلّما* قرب منها من القطبين كان أصغر وكلّما بعد عن القطبين كان أعظم. وأعظم هذه الدوائر هي التي تمرّ بمركز العالم وبعدها من القطبين بعد واحد وهي تفصّل العالم بنصفين لأنّها تمرّ بمركزه. [٥٢] وهذه الدائرة تسمّي* دائرة معدّل النهار لأنّ الشمس إذا انتهت إليها اعتدل الليل والنهار في جميع العالم* وقطبا هذه الدائرة هما* قطبا العالم اللذين سمّيناهما الشمالي والجنوبي. [٥٣] والذي نسمّيه قطب الدائرة هي النقطة التي تكون كلّ الخطوط الخارجة منها إلى محيط تلك الدائرة* متساوية وباقي الدوائر المتوازية تسمّي دوائر الأزمان.

[٥٤] وهذه الدوائر وكلّ الدوائر التي في الفلك تقسّم بثلاثمئة وستّين أجزاء* متساوية قسمةً متوهّمةً وكلّ جزء منها يسمّي درجة. فالدوائر الزمانية تسمّي هذه الأجزاء منها أجزاء الزمان. [٥٥] وتقسم أيضا هذه الدوائر بأربعة وعشرين قسماً كلّ قسم منها هو خمسة عشر* درجة يسمّي* ساعة فتكون* الدائرة كلّها أربعة وعشرين ساعةً.

[٥٦] فإذا دار الفلك الأعظم فحرّك جميع الكواكب تحرّك كلّ كوكب

منها على دائرة من هذه الدوائر فكلّما قطعت نقطة من الفلك خمسة عشر درجةً قيل قد مضت ساعة واحدة فإذا قطعت جميع الدائرة يكون الماضي من الزمان أربع وعشرون ساعةً. [٥٧] وهي تقطع الدائرة بدورة واحدة يدورها الفلك الأعظم وبدورة* هذا الفلك التامة يكون اليوم والليلة فلذلك صار اليوم والليلة أربع وعشرون ساعةً.

[٥٨] ونتوهّم* أيضاً دائرةً عظيمةً تمرّ بمركز العالم وبالقطبين فتقطع* دائرة معدّل النهار بنصفين وتقطع* العالم بأسره وتقطع* كرة الأرض (١٠٣١/ي) . [٥٩] وتقطع* أيضاً دائرة معدّل النهار كرة الأرض بنصفين فيقسم الأرض بهاتين الدائرتين بأربعة أقسام متساوية قسمان منها مما يلي الشمال وقسمان منها مما يلي الجنوب. [٦٠] وأحد القسمين الشماليين هو الربع الغاصّ من الأرض . والدائرة التي تحدث* في سطح الأرض من فصل معدّل النهار تسمّي* خطّ الاستواء* والنصف من هذه الدائرة الذي* يحدّ المعمورة من الأرض هو طول العمارة وهو مائة وثمانون جزءاً من أجزاء الدائرة. [٦١] وفي وسط هذا الطول على خطّ الاستواء موضع يسمّي وسط الأرض ويسمّي القبّة.

[٦٢] والدائرة الأخرى المارّة على القطبين الفاصلة للأرض تسمّي أفقاً لوسط الأرض وتسمّي* أفقاً لخطّ الاستواء. وهذه الدائرة تفصّل العالم بنصفين فيكون أحد النصفين ظاهرا عند (٨١/ك) من يكون في وسط الأرض والنصف الآخر خفياً عنده. [٦٣] وكلّ دائرة عظيمة تقسم* العالم بنصفين وتحدّ* بين ما تظهر وما لا تظهر تسمّي* أفقاً والموضع الذي في وسط الأرض هو كالقطب للدائرة التي تمرّ بالقطبين وتحدّ العمارة. [٦٤] والخطّ الخارج من مركز الأرض إلى ذلك القطب إذا خرج على استقامة حتّى ينتهي إلى الفلك الأعلى* فإنّ النقطة التي ينتهي إليها من سطح الفلك الأعلى تسمّي سمت الرأس لأنّها تسامت رؤوس* الذين يكونون في وسط الأرض وتلك النقطة هي قطب بالحقيقة للدائرة العظيمة التي هي أفق لوسط الأرض. [٦٥] وكلّ نقطة من الأرض فإنّ النقطة المسامتة لها من الفلك الأعلى تسمّي سمت الرأس لذلك الموضع من الأرض. والدائرة العظيمة التي تلك النقطة قطب لها هي دائرة الأفق لذلك الموضع من الأرض*.

[٦٦] ولكلّ* نقطة* من الأرض أفق يفصّل العالم بنصفين. [٦٧] فأما آفاق النقط التي على خطّ الاستواء فيكون* قطبا العالم على محيطها وتكون* دائرة معدّل النهار قائمةً (ب٧/ك) على كلّ واحد منها قياماً معتدلاً*. [٦٨] وأما باقي الآفاق فيكون أحد القطبين فيها ظاهراً والآخر خفياً وتكون* دائرة معدّل النهار مائلةً عليها* أما آفاق المساكن الشمالية فيكون معدّل النهار مائلا* عليها إلى* ما يلي الجنوب ويكون القطب الشمالي فيها ظاهراً فوق الأرض وأما الآفاق الجنوبية فيكون معدّل النهار مائلاً* عليها إلى ما يلي الشمال فيكون القطب الجنوبي فيها ظاهراً فوق الأرض.

[٦٩] والدائرة العظيمة التي نتوهّم مارّةً بقطبي العالم وبنقطة سمت الرأس في أي* الآفاق كانت تسمّي دائرة نصف النهار لذلك الأفق* لأنّ الشمس إذا انتهت إليها بالحركة السريعة اليومية انتصف النهار في ذلك الموضع وكذلك إذا انتهت إليها من النصف الذي* تحت الأفق انتصف الليل. [٧٠] وهذه الدائرة تقسم* الأفق بنصفين فتحدث* من تقاطعها نقطتان أحديهما مما يلي الشمال وتسمّي النقطة الشمالية والأخرى* مما يلي الجنوب وتسمّي* (٩١/ك) النقطة الجنوبية. [٧١] وكذلك دائرة معدّل النهار تقسم* الأفق بنصفين فتحدث* نقطتين آخرتين تسمّي* أحديهما نقطة المشرق والأخرى نقطة المغرب.

[٧٢] ففي كلّ أفق من الآفاق أربع نقط هي الشمال والجنوب والمشرق والمغرب* والخطّ المستقيم الذي يصل بين نقطتي المشرق والمغرب يقال له خطّ المشرق والمغرب*. [٧٣] والخطّ المستقيم الذي يصل بين نقطتي الشمال والجنوب يقال له خطّ الشمال والجنوب وهذا الخطّ هو الذي يسمّي أيضاً خطّ نصف النهار وهو الذي يستخرج في سطوح الرخامات وهو الفصل المشترك بين دائرة نصف النهار وبين سطح دائرة الأفق.

[٧٤] ونتوهّم* أيضاً دائرةً عظيمةً تمرّ بسمت الرأس وهو قطب الأفق وبمركز العالم تكون* قائماً على الأفق على زويا قائمة وتقطع* محيط الأفق على نقطتين وتقطع* سطح الأفق على خطّ مستقيم يصل بين تينك النقطين وتمرّ بمركز العالم وهذه الدائرة تسمّي الدائرة السمتية (ب١٠٢/ي). [٧٥] وكذلك كلّ الدوائر التي على هذه الصفة (ب٩/ك)

تسمّي* الدوائر السمتية وتانك النقطتان* تسمّي كلّ واحدة منها نقطة السمت ويسمّي* الخطّ المستقيم خطّ السمت. [٧٦] وتسمّي القوس من الأفق التي بين نقطة الشمال والجنوب وبين النقطة السمتية قوس السمت وربّما تسمّي أيضاً القوس التي بين نقطة المشرق أو* المغرب وبين النقطة السمتية سمتاً.

القول على الأطوال والعروض

[٧٧] عرض المدينة هو قوس من دائرة نصف النهار فيما بين سمت الرأس وبين دائرة معدّل النهار. [٧٨] وتسامتها* على سطح الأرض قوس من دائرة عظيمة محيطة بكرة الأرض مركزها مركز الأرض فيما بين أيّ نقطة فُرضت في تلك المدينة وبين خطّ الاستواء الذي يحدّه سطح معدّل النهار. [٧٩] وفي سطح هذه الدائرة يكون خطّ نصف النهار المستخرج في سطح أفق تلك المدينة وهذه القوس هي بعد المدينة من خطّ الاستواء.

[٨٠] فأما طول المدينة فهو قوس من دائرة معدّل النهار فيما بين دائرة نصف نهار تلك المدينة وبين آخر العمارة أما من جهة المشرق (١٠١/ك) وأما من جهة المغرب وهذه الدائرة هي قطعة من نصف دائرة معدّل النهار الذي ذكرنا أنّه يحدّ العمارة. [٨١] وطرفا* العمارة هما نقطتا تقاطع هذا النصف لدائرة الأفق أعني أفق وسط الأرض.

[٨٢] وتتوهّم* دوائر موازية لمعدّل النهار من الدوائر الزمانية تقطع* العالم وتقطع آفاق المساكن كلّ دائرة منها تقطع* كلّ افق على نقطتين ويقطعها الأفق بقسمين* أما آفاق خطّ الاستواء فإنّها تقطع هذه الدوائر كلّها بنصفين لأنّ هذه الآفاق تمرّ بقطبي هذه الدوائر. [٨٣] فأما باقي الآفاق فإنّها تقسم دائرة معدّل النهار فقط بنصفين* وتقسم* هذه الدوائر بأقسام مختلفة تكون* الأقسام التي فوق الأفق غير مساوية للأقسام التي تحت الأفق وكلّما بعد عن معدّل النهار إلى جهة القطب الظاهر كان القسم الذي فوق الأرض أعظم من الشبيه بما قرب من معدّل النهار وكلّما بعد عن معدّل النهار إلى جهة القطب الخفي كان القسم الذي فوق الأفق* أصغر من (ب١٠/ك) الشبيه بما قرب من معدّل النهار.

[٨٤] وباختلاف هذه القطع يختلف الليل واليوم في الآفاق المائلة وهذه القسي* هي التي تسمّي قسي* النهار وكلّ كوكب من الكواكب يتحرّك على دائرة من هذه الدوائر فالقوس الأعلى من هذه الدوائر تسمّي*

قوس النهار لذلك الكوكب والقوس السفلى منها تسمّي* قوس الليل. [٨٥] فإذا تحرّك الشمس على دائرة من هذه الدوائر كانت* أجزاء القوس العليا من هذه الدائرة هي أجزاء ساعات النهار وأجزاء القوس السفلى هي أجزاء ساعات الليل. [٨٦] ولما كانت هذه القسي كلّما بعدت عن معدّل النهار إلى جهة القطب الظاهر كان ما يظهر منها أكثر شبهاً مما قرب وأعني بأكثر شبهاً* هو الذي أجزاؤه المقسومة* بها الدائرة بثلاثمئة وستّين جزء أكثر عدداً وجب أن تكون* أجزاء ما بعد من هذه القسي أكثر عدداً* ولذلك لما كانت* أجزاء ما بعد من هذه القسي من معدّل النهار أكثر عدداً كانت الشمس إذا تحرّكت على هذه الدوائر* التي هي أبعد كان مكثها فوق* الأرض أكثر وكانت* أجزاء ساعات النهار أكثر.

[٨٧] (١١١/ك) وآخر دائرة تتحرّك* عليها الشمس من هذه الدوائر مما يلي القطب الشمالي تسمّي مدار السرطان وأطول ما يكون النهار في الآفاق الشمالية إذا تحرّكت الشمس على هذه الدائرة. [٨٨] ويلزم عكس ذلك في جميع الدوائر التي تلي* القطب الجنوبي أعني أن القوس* التي فوق الأفق كلّما بعد منها من معدّل النهار كان أقلّ شبهاً مما قرب وكذلك كلّما بعدت الشمس في جهة الجنوب كان النهار أقصر وآخر دائرة تتحرّك* عليها الشمس من هذه الدوائر تسمّي مدار الجدي.

[٨٩] وأما آفاق خطّ الاستواء فليس يعرض فيها شيء من ذلك بل قوس النهار لجميع الكواكب (١٠٤١/ي) مثل قوس الليل لأنّ الآفاق تقسم* جميع الدوائر المتوازية لمعدّل النهار بنصفين فتكون* القوس التي فوق الأفق مساويةً للقوس* التي تحت الأفق* ولذلك يستوي الليل ولنهار أبداً في هذه المواضع لأنّ مكث الشمس فوق الأفق* يكون مساوياً لمكثها تحت الأفق .

[٩٠] وكلّ دائرة من الدوائر المتوازية التي تقطع* الآفاق فهي تقطعها* على نقطتين (ب١١/ك) أحديهما* مما يلي المشرق والأخرى مما يلي المغرب. [٩١] فكلّ كوكب يتحرّك على تلك الدائرة فهو يطلع من النقطة الشرقية ويغرب من النقطة الغربية المقابلة لها وكذلك كلّ نقطة على سطح الفلك الأعلى تطلع* من نقطة من هذه النقط* وتغرب* من النقطة الغربية* المقابلة لها. [٩٢] والقوس من الأفق التي بين النقطة الشرقية وبين نقطة

المشرق من معدّل النهار تسمّي* سعة المشرق لذلك الكوكب أو لتلك* النقطة التي تطلع* من النقطة الشرقية من الأفق. والقوس من الأفق التي فيما بين النقطة الغربية وبين نقطة المغرب من معدّل النهار تسمّي* سعة المغرب.

[٩٣] فقد ذكرنا أن الدوائر المتوازية كلّما قربت من القطب الظاهر عظمت القوس التي فوق الأفق ولا يزال كذلك إلى أن ينتهي إلى دائرة تماس الأفق على نقطة واحدة فيكون جميع الدائرة ظاهراً فوق الأفق. [٩٤] فكلّ كوكب يتحرّك على هذه الدائرة يكون دائم الظهور ويلزم عكس ذلك مما يلي القطب الخفي حتّى ينتهي إلى دائرة تماس الأفق على (١٢١/ك) نقطة واحدة فيكون جميعها تحت الأفق فيكون الكوكب الذي يتحرّك عليها أبداً خفياً.

[٩٥] وما جاوز هاتين الدائرتين أما ما* يلي القطب الظاهر فيكون جميعها ظاهراً* وتكون* مرتفعةً عن الأفق وأما ما يلي القطب الخفي* فيكون جميعها خفياً* وتكون* منخفضةً عن الأفق•

[٩٦] وجميع المعمور من الأرض على ما ذكر فيما بين دائرتين متوازيتين أحديهما دائرة معدّل النهار وتمرّ بسمت رؤوس من يكون على خطّ الاستواء والأخرى الدائرة التي بعدها من معدّل النهار ستّة وستّون جزءاً من الأجزاء التي بها الدائرة ثلاثمئة وستّون جزءاً. وتمرّ هذه الدائرة بسمت رؤوس من يكون في آخر العمارة وفي هذا المكان يكون مدار رأس السرطان مماساً للأفق.

[٩٧] وتقسم* هذه القطعة من الأرض بسبعة أقسام تسمّي* سبعة أقاليم• [٩٨] أما ابتداء الأقليم* الأوّل منها فخطّ* الاستواء والليل والنهار في هذا الموضع أبداً* مساويتين ثمّ لا يزال الليل والنهار يختلفان (ب١٢/ك) عند كلّ من قرب من الشمال كما ذكرنا آنفاً حتّى ينتهي إلى آخر الأقليم* السابع وهو آخر العمارة. [٩٩] وفي هذا المكان أطول ما يكون النهار أربعة وعشررون ساعةً وذلك أنّ مدار* السرطان يماس أفقهم على نقطة واحدة فيكون جميعه* ظاهراً فوق أفقهم فإذا انتهت الشمس إلى هذه الدائرة كانت في الدورة الكاملة فوق الأفق فلذلك يكون أطول نهارهم

أربعة وعشرين ساعةً ، ثمّ ترجع الشمس ولا يزال النهار يتناقص والليل يتزايد إلى أن تنتهي إلى* أبعد بعدها في الجنوب فإذا صارت إلى مدار الجدي كانت في الدورة الكاملة تحت الأفق لأنّ مدار* الجدي يماس الأفق على نقطة واحدة ويكون جميعه تحت الأفق ولذلك يكون أطول ليلهم أربعة وعشرين ساعةً. [١٠٠] وعرض هذا المكان وهو القوس من دائرة نصف النهار فيما بين سمت الرأس وبين معدّل النهار قريب من ستّة وستّين جزءاً من الأجزاء التي بها الدائرة ثلاثمئة وستّون جزءاً ويقال إنّه ليس (١٣١/ك) فيما بعد هذه الدائرة* شيء من العمارة.

[١٠١] وقد تقسم* الأقاليم السبعة قسمةً غير هذه فمنهم ما يجعل مبدأ العمارة* (ب١٠٣/ي) دون خطّ الاستواء ونهايتها دون عرض ستّة وستّين ولا يعتدّ بما عن جنبتي ذلك من العمارة*. [١٠٢] فأما بطلميوس فإنّه يجعل نهاية العمارة في الشمال عرض ثلاثة وستّين ويجعل* نهايتها في الجنوب* متجاوزاً لخطّ الاستواء* إلى عرض ستّة عشر* وربُع وسُدس* على ما ذكر في كتابه في المعمورة.

[١٠٣] فأما ما تجاوز عرض ستّة وستّين من الأرض فإنّه كلّما قرب من الشمال ارتفعت دائرة مدار السرطان عن الأفق فصار الأفق مماساً لدائرة أعظم منها. [١٠٤] فما دامت الشمس على دائرة من الدوائر المتوازية التي فيما بين الدائرة الماسة ومدار السرطان فإنّه يكون النهار* في هذا المكان لأنّ الشمس تكون ظاهرةً فوق الأرض ومثل ذلك يكون ليلاً إذا كانت الشمس مما يلي القطب الجنوبي*.

[١٠٥] ولا يزال الليل والنهار* (ب١٢/ك) يتزايد كلّما قرب المكان من القطب الظاهر إلى أن ينتهي إلى المكان السامت للقطب وهو الذي يكون قطب العالم عنده على سمت الرأس ويكون محور العالم منتصباً فعند ذلك يكون الأفق دائرة معدّل النهار. [١٠٦] وتكون* سنة الشمس في هذا المكان يوماً وليلةً يكون النهار قريباً من ستّة أشهر وذلك ما دامت الشمس في البروج الشمالية لأنّها تكون* أبداً فوق الأفق* ويكون الليل أيضا قريباً من

ستّة أشهر وذلك ما دامت الشمس في البروج الجنوبية لأنّها تكون* أبدأ تحت الأفق والشمس في هذا المكان تتحرّك* أبدأ حول الأفق ولا ترتفع* ارتفاعاً كثيراً وأكثر ما تكون* ارتفاعاً إذا كانت على مدار السرطان.

القول على فلك البروج

[١٠٧] قد قدّمنا أن لكلّ كوكب من الكواكب كرة تخصّه وأن لكلّ كرة منها حركة تخصّها وأن كلّ كوكب منها ترسم نقطة مركزه بحركة كرته دائرة متوهّمة في كرته (١٤١/ك) فإذا تحرّك فلك الشمس من المغرب إلى المشرق حدثت* من حركة مركز الشمس دائرة عظيمة في فلك الشمس.
[١٠٨] ونتوهّم* هذه الدائرة قاطعةً للعالم فتحدث* في سطح الفلك الأعلى دائرة عظيمة مركزها مركز العالم وهذه الدائرة هي التي تسمّي* فلك البروج وقطبا هذه الدائرة يسمّيان* قطبا فلك البروج . [١٠٩] وهذه الدائرة تقطع* دائرة معدّل النهار بنصفين على نقطتين متقابلتين تسمّي أحديهما نقطة الاعتدال الربيعي وتسمّي الأخرى نقطة الاعتدال الخريفي.
[١١٠] ونتوهّم* أيضاً دائرة عظيمة تمرّ بقطبي معدّل النهار وهما قطبا العالم وبقطبي فلك البروج فتقطع* دائرة البروج على نقطتين متقابلتين أحديهما مما يلي الشمال والأخرى مما يلي الجنوب أما الشمالية فتسمّي نقطة الانقلاب الصيفي وأما الجنوبية فتسمّي* نقطة الانقلاب الشتوي.
[١١١] ونتوهّم* أيضاً دائرة عظيمة تمرّ بقطبي دائرة البروج وبقطبي الاعتدالين فهاتان الدائرتان تقسمان* دائرة البروج بأربعة أقسام متساوية أما الربع الذي بين نقطة الاعتدال الربيعي وبين نقطة الانقلاب الصيفي فهو الذي يحدّ زمان الربيع وذلك ما دامت الشمس بحركة فلكها الخاصّ مسامةً لهذه القوس وهو الزمان الذي يسمّي الربيع.
[١١٢] وأما الربعّ الذي بين بقطة الانقلاب الصيفي وبين نقطة الاعتدال الخريفي فهو الذي يحدّ زمان الصيف وذلك ما دامت الشمس مسامتةً بحركتها هذه القوس. [١١٣] وأما الربع* الذي بين نقطة الاعتدال الخريفي وبين نقطة الانقلاب الشتوي فهو الذي يحدّ زمان الخريف* وذلك ما دامت الشمس مسامتةً له بحركتها. [١١٤] فأما الربع* الذي (ب١٤/ك) فيما بين نقطة الانقلاب الشتوي وبين نقطة الاعتدال الربيعي فهو الذي يحدّ

زمان الشتآء وذلك ما دامت الشمس تسامته* بحركتها وجميع هذه الدائرة تحدّ زمان السنة الشمسية*.

[١١٥] وإذا تحرّكت الشمس بحركتها الخاصّة* من مسامتة نقطة من هذه الدائرة (١٠٥١/ي) فقطعت جميع الدائرة حتّى تعود* إلى مسامتة تلك النقطة فتلك الحركة هي سنة الشمس الكاملة. [١١٦] وأوّل السنة الشمسية هو وقت مسامتة الشمس لنقطة الاعتدال الربيعي فإذا تحرّكت متوجّهةً إلى الشمال يكون زمان الربيع وكلّما قربت من نقطة الانقلاب الصيفي سخن الزمان عند أهل المساكن الشمالية لقرب الشمس من مسامتة رؤوسهم ولما قدّمنا من طول مكث الشمس فوق الأفق. فلا يزال كذلك إلى أن تنتهي* إلى نقطة الانقلاب الصيفي فقد اشتدّ الحرّ لتواتر اسجان الشمس الهواء. [١١٧] ثمّ لا تزال* كذلك متوجّهةً إلى نقطة الاعتدال الخريفي فيعرض في هذه الربع من سخونة الهواء أكثر مما عرض في الربع الأوّل لطول تواتر الاسخان حتّى تنتهي* إلى نقطة الاعتدال الخريفي فعندها يعود الزمان معتدلاً.

[١١٨] ثمّ تتحرّك* متوجّهةً إلى الجنوب فيبرد الهواء عند أهل المساكن الشمالية وذلك لبعد الشمس عن مسامتة رؤوسهم فكلّما بعدت في الجنوب أزاد برد الهواء إلى أن تنتهي* إلى نقطة الانقلاب الشتوي فعندها تشتدّ البرد لأنّها في هذا المكان أبعد ما تكون* من الشمال ولتواتر تبريدها* بالعرض للهواء. [١١٩] ثمّ تعود* متوجّهةً إلى نقطة الاعتدال الربيعي فيعرض* في هذا الربع أيضاً من شدّة البرد أكثر مما عرض في الربع الذي قبله لمثل ما ذكرنا في الربع الصيفي من طول تواتر تبريد الهواء. فإذا انتهت إلى النقطة الربيعية فهو آخر سنتها (١٥١/ك).

[١٢٠] ونتوهّم* أيضاً دائرتين عظيمتين* تخارجان* من قطبي دائرة البروج فتقطعان* الربع الربيعي من دائرة البروج بثلاثة* أقسام متساوية وتقطعان* أيضاً الربع الخريفي المقابل لهذا الربع بثلاثة أقسام متساوية. ونتوهّم* أيضاً دائرتين آخريين* تخرجان من قطبي دائرة البروج وتقطعان* الربع الصيفي* والربع الشتوي المقابل له كلّ واحد منهما بثلاثة أقسام متساوية. [١٢١] فيصير جملة الدوائر الخارجة من قطبي دائرة البروج ستّ* دوائر تقسم* دائرة البروج بإثنى عشر قسماً متساويةً

وتقسم* الفلك الأعلى بإثنى عشر قسماً متساويةً كلّ قسم منها يحدّه نصفا دائرتين من هذه الدوائر الستّ* تسمّي هذه الأقسام الإثنى عشر من الفلك الأعلى بروجاً وتسمّي* الأقسام الإثنى عشر من دائرة البروج بروجا. [١٢٢] وكلّ واحد من هذه القسي أعني أقسام دائرة البروج مقسومة بثلاثين قسماً متساويةً قسمةً متوهّمةً يسمّي كلّ جزء منها درجة فالدائرة بجملتها ثلاثمئة وستّون درجة.

[١٢٣] ونتوهّم* هذه الدوائر الستّ* قاطعةً* للعالم فهي تقسم جميع أفلاك* الكواكب بإثنى عشر قسماً متساويةً شبيهةً بأقسام الفلك الأعلى تسمّي أيضاً بروجاً (ب١٥/ك). [١٢٤] وتقسم* دائرة البروج أفلاك الكواكب أيضاً وتحدث* في كلّ واحد منها دائرة موازية للدائرة العظمى تسمّي كلّ واحدة منها الفلك المثّل بفلك البروج. [١٢٥] وتنقسم* أيضاً هذه الدوائر بالدوائر الستّ* بإثنى عشر قسماً متساويةً تسمّي كلّ قوس منها برجاً. وينقسم أيضاً فلك الكواكب الثابتة بالدوائر الستّ بإثنى عشر قسماً متساويةً تسمّي* أيضاً بروجاً*.

[١٢٦] وفي فلك الكواكب الثابتة كواكب كثيرة متشكّلة بأشكال مختلفة ففي أحد الأقسام الإثنى عشر كواكب من جملة ما يشتمل عليه من الكواكب الثابتة متشكّلة بشكل شبيه بصورة الحمل فيسمّي* هذا البرج* من الفلك برج الحمل وتسمّي* القوس من دائرة البروج التي في وسط هذا القسم أيضاً برج الحمل. [١٢٧] وهذا البرج من الفلك هو البرج الأوّل من أقسام* الربع الربيعي وهو بالفرض أوّل دائرة البروج وابتداء هذه* القوس (ب١٠٥/ي) هو نقطة الاعتدال الربيعي* إذ* هي بالفرض أوّل الدائرة وهي رأس الحمل.

[١٢٨] وفي القطعة* التي تلي هذه القطعة من الفلك كواكب متشكّلة بشكل شبيه بصورة الثور فيسمّي* هذا البرج (١٦١/ك) برج الثور وتسمّي* القوس من الدائرة التي فيه أيضاً برج الثور. [١٢٩] وكذلك جميع البروج في كلّ واحد منها صورة يسمّي البرج باسم تلك الصورة وتسمّي القوس من دائرة البروج التي في ذلك البرج باسم ذلك البرج.

[١٣٠] والبرج الثالث التؤمان والرابع السرطان وهو أوّل البروج*

الصيفية وأوّله هو نقطة الانقلاب الصيفي والخامس الأسد والسادس السنبلة والسابع الميزان وهو أوّل البروج الخريفية وأوّله هو نقطة الاعتدال الخريفي والثامن العقرب والتاسع القوس والعاشر الجدي وهو أوّل البروج الشتوية وأوّله هو نقطة الانقلاب الشتوي والحادي عشر الدلو والثاني عشر الحوت وهو آخر البروج وآخره نقطة الاعتدال الربيعي .

[١٣١] وإذا تحرّكت الشمس بحركتها الخاصّة فهي تكون* أبداً مسامتةً لدائرة البروج فإذا توهّمنا خطّاً مستقيماً يخرج من مرجز العالم وينتهي إلى مركز الشمس ويخرج على استقامة إلى أن ينتهي إلى الفلك الأعلى فهو ينتهي إلى نقطة من دائرة البروج فتلك النقطة هي النقطة التي نقول إنّ (ب١٦/ك) الشمس نازلة فيها في ذلك الوقت من البرج* والدرجة التي فيها تلك النقطة.

[١٣٢] فأما سائر الكواكب فإنّ نزولها* في سائر البروج كما أصف نتوهّم خطّاً مستقيماً يخرج من مركز العالم إلى مركز الكوكب ويخرج* على استقامته حتّى ينتهي إلى سطح الفلك الأعلى فيقع في برج من البروج الإثنى عشر التي ينقسم بها جميع سطح الفلك فنقول إنّ الكوكب في ذلك الوقت في ذلك البرج.

[١٣٣] ثمّ نتوهّم دائرةً عظيمةً تخرج من قطبا دائرة البروج وتمرّ بالنقطة التي إليها انتهى ذلك الخطّ مارّةً* إلى القطب الآخر فهي تقطع* القوس من دائرة البروج التي في وسط ذلك البرج فنقول إنّ الكوكب في تلك الدرجة التي قطعتها تلك الدائرة في ذلك البرج وهذا هو موضع الكوكب في الطول*. [١٣٤] وكذلك إن انتهي طرف الخطّ إلى محيط واحدة* من الدوائر الستّ* فذلك هو موضع الكوكب والنقطة التي تقطع* عليها تلك* الدائرة دائرة البروج هي موضعه من دائرة البروج وهي أحد رؤوس البروج.

[١٣٥] فأما عرض الكوكب فهو القوس من الدائرة المارّة بقطبي (١٧ا/ك) دائرة البروج التي فيما بين النقطة الأولى التي هي طرف الخطّ المارّ بمركز الكوكب وبين دائرة البروج هكذا موضع جميع الكواكب السيارة والثابتة في الطول والعرض.

[١٣٦] ونتوهّم أيضاً دوائر موازية لمعدّل النهار تقطع* دائرة

البروج على النقط التي هي رؤوس البروج وتسمّي* هذه الدوائر مدارات البروج لأنّ نقط رؤوس البروج تدور على هذه الدوائر وأوّلها مما يلي الشمال مدار* السرطان وهي تماس دائرة البروج على نقطة واحدة وهي رأس السرطان وهي نقطة الانقلاب الصيفي . [١٣٧] والدائرة الثانية هي التي تمرّ برأس الجوزاء وبرأس* الأسد والثالثة تمرّ برأس الثور وبرأس* السنبلة والرابعة هي دائرة معدّل النهار وتمرّ برأس الحمل وبرأس* الميزان والخامسة تمرّ برأس الحوت وبرأس العقوب والسادسة تمرّ برأس الدلو وبرأس القوس والسابعة هي مدار الجدي وتماس دائرة البروج على نقطة واحدة وهي رأس الجدي وهي نقطة الانقلاب الشتوي.

[١٣٨] وهذه المدارات تقطع* الآفاق وتحصل* بين كلّ دائرتين من مدارات البروج قوس من الأفق تسمّي* (ب١٧/ك) تلك* القوس سعة المشرق لذلك البرج (١٠٦١/ي) الذي فيما بين الدائرتين لأنّ الفلك إذا تحرّك* بكليته حرّك جميع ما يشتمل عليه وهو يحرّك دائرة البروج فكلّ نقطة منها تطلع* من نقطة من الأفق فيما يلي المشرق وتغرب من نقطة منها* مما يلي المغرب وتتحرّك تلك النقطة على دائرة موازية لمعدّل النهار. [١٣٩] فنقطة رأس الحمل تتحرّك على دائرة معدّل النهار ونقطة رأس الثور تتحرّك على مدار الثور وكلّ نقطة من برج الحمل وبرج السنبلة تتحرّك على دائرة* فيما بين هاتين الدائرتين وتطلع من النقطة التي تقطع* عليها هذه الدائرة محيط الأفق فكلّ نقطة من برج الحمل وبرج السنبلة تطلع* من نقطة من النقط التي على القوس من الأفق التي فيما بين مدار الحمل ومدار الثور فلذلك سمّيت هذه القوس سعة مشرق الحمل وكذلك كلّ قوس من القسي الباقية التي فيما بين الدوائر المتوازية تسمّي* كلّ واحدة منها سعة المشرق لذلك البرج وللبرج أيضاً الذي معه فيما بين الدائرتين المتوازيتين.

القول على الميل

[١٤٠] الميل هو قوس من دائرة عظيمة تمرّ بقطبي العالم فيما بين (١٨١/ك) دائرة البروج وبين دائرة معدّل النهار. [١٤١] وكلّ نقطة من دائرة البروج لها ميل عن دائرة معدّل النهار سوى نقطتي التقاطع اللتين هما رأسا الحمل والميزان. [١٤٢] وذلك أنّ كلّ الدوائر التي تخرج* من قطبي العالم إلى كلّ نقطة من دائرة البروج فانّها تقطعها* وتنتهي* إلى دائرة معدّل النهار فتحصل* بين كلّ نقطة من دائرة البروج وبين معدّل النهار قوس من الدائرة التي تمرّ بالقطبين فتلك القوس هي ميل تلك النقطة.

[١٤٣] وهذه الميول مختلفة وكلّما* قرب منها من نقطتي التقاطع من دائرة البروج ودائرة معدّل النهار أعني نقطتي الاعتدال كان أصغر وكلّما* بعد منها عن هاتين النقطتين كان أعظم . [١٤٤] فأعظم* الميول ميل النقطة التي في وسط النصف من دائرة البروج وهي نقطة الانقلاب* فالميل الأعظم هو قوس من الدائرة التي تمرّ بقطبي العالم وبنقطتي الانقلابين التي فيما بين نقطة الانقلاب وبين دائرة معدّل النهار وهذه القوس هي قريبة* من أربعة وعشرين جزءاً من الأجزاء التي بها الدائرة كلّها ثلاثمئة وستّون جزءاً.

[١٤٥] فأما (ب١٨/ك) القسي من الدوائر التي تمرّ بالنقط التي تحدّ البروج فانّها تسمّي ميول تلك البروج أما التي تمرّ برأس الثور وبرأس* العقرب فإنّ القوس منها التي فيما بين رأس الثور وبين معدّل النهار تسمّي* ميل الحمل والمقابلة* لها تسمّي ميل الميزان . [١٤٦] وأما التي تمرّ برأس الجوزاء وبرأس القوس فإنّ القوس منها* التي هي الميل تسمّي* ميل الثور والعقرب*. [١٤٧] وأما التي تمرّ برأس الأسد وبرأس الدلو فإنّها تسمّي ميل السرطان والجدي*. [١٤٨] وأما التي تمرّ برأس السنبلة وبرأس* الحوت فإنّها تسمّي ميل الأسد والدلو*. [١٤٩] وأما التي تمرّ بالنقط التي فيما بين هذه النقط فإنّها تسمّي ميل تلك الدرجة أعني

التي تحدّها تلك النقطة وكلّ هذه الميول مساوية لميول النقط المقابلة لها.

[١٥٠] وكلّ نقطة من دائرة البروج فإنّها تتحرّك كما قدّمنا على دائرة* موازية لمعدّل النهار وبعد تلك الدائرة من معدّل النهار هو ميل تلك النقطة وإذا كانت الشمس مسامتةً لنقطة من دائرة البروج فإنّها تتحرّك في سطح الدائرة الموازية لمعدّل النهار التي ترسمها تلك (١٩١/ك) النقطة فيكون ميلها عن معدّل النهار أبداً هو ميل النقطة من دائرة البروج التي تسامتها.

[١٥١] ولأنّا قد بيّنا أن عروض المساكن هي قسي من الدوائر التي تمرّ بقطبي العالم وبنقط سمت الرأس فيما بين سمت الرأس وبين معدّل النهار فيجب من ذلك أن تكون* دوائر الميول تنطبق بحركة الكلّ على دوائر (ب١٠٦/ي) العرض فكلّ نقطة من دائرة البروج يكون ميلها مساوياً لعرض من العروض فإنّ تلك النقطة بحركة الكلّ تنطبق على نقطة سمت الرأس لذلك العرض في كلّ دورة. [١٥٢] ولذلك إذا كانت الشمس مسامتةً لتلك النقطة من دائرة البروج فإنّها تمرّ في ذلك اليوم بسمت الرأس لكلّ من تحت ذلك العرض فيجب من ذلك أن تكون* الشمس تمرّ بسمت رؤوس أهل المساكن الذين عرضهم ليس بأكثر من الميل الأعظم. [١٥٣] أما الذين عرض مسكنهم مساو للميل الأعظم وهو قريب من أربعة وعشرين جزءاً* فإنّ الشمس تمرّ بسمت رؤوسهم إذا صارت (ب١٩/ك) في نقطة الانقلاب* وحينئذ لا يكون لأشخاص المقاييس ظلّ في نصف نهار هذا* اليوم عندهم . فأما باقي السنة فيكون ظلّ أشخاص المقاييس إلى ما يلي الشمال عن ذلك المساكن.

[١٥٤] وأما المساكن التي عرضها أقلّ من الميل الأعظم فإنّ الشمس تمرّ بسمت رؤوسهم في كلّ سنة دفعتين لأنّ الدوائر الموازية لمعدّل النهار التي تقطع* دائرة البروج فإنّها تقطعها* على نقطتين وميل هاتين النقطتين متساو* لأنّه* بعد تلك الدائرة عن معدّل النهار فإذا انتهت الشمس إلى النقطة الأولى من تينك النقطتين تحرّكت* في سطح تلك الدائرة فمرّت بسمت رؤوسهم الذين عرضهم مساو لميل تلك الدرجة ولا يكون حينئذ لأشخاص مقاييسهم ظلّ في نصف النهار ذلك اليوم. [١٥٥] فإذا تحرّكت

الشمس من تلك النقطة إلى جهة الشمال زالت عن سمت رؤوسهم وصارت أظلال مقاييسهم إلى جهة (٢٠١/ك) الجنوب فلا يزال كذلك إلى أن تنتهي* إلى النقطة الثانية* فتمرّ* حينئذ أيضاً على سمت رؤوسهم فإذا تحرّكت من تلك النقطة إلى جهة الجنوب صارت أظلال مقاييسهم إلى جهة الشمال.

[١٥٦] وأما الذين* عرضهم أكثر من الميل الأعظم فليست* تمرّ الشمس بسمت رؤوسهم بل تكون* أبداً مائلةً إلى الجنوب عنهم ويكون أظلال مقاييسهم أبداً إلى جهة الشمال وهذا الذي ذكرنا هو حال ميول الشمس.

[١٥٧] فأما ميل* الكواكب السيارة والثابتة فهو قوس من دائرة عظيمة فيما بين النقطة المسامتة لمركز الكوكب وبين معدّل النهار أما من الدوائر التي تخرج من قطبي العالم وتسمّي هذه* القوس ميل الكوكب وأما من الدوائر التي تخرج* من قطبي دائرة البروج فربّما سمّيت القوس ميل الكوكب وربّما سمّيت عرض الكوكب بالقياس إلى معدّل النهار*

[١٥٨] وتختلف ميول الكواكب السيارة كما يختلف ميل الشمس لأنّ لكلّ واحد منها دائرة تخصّه مقاطعة لدائرة* معدّل النهار تكون* حركته* بحسبها فيميل إلى الشمال وإلى الجنوب كما تميل الشمس إلى* هاتين الجهتين (ب٢٠/ك) وكلّما بعد أيضاً من نقطتي التقاطع كان أكثر ميلاً.

[١٥٩] فأما الكواكب الثابتة فإنّ ميلها ليس يختلف اختلافاً بيّناً لأنّها بطيئة الحركة لكن في الزمان الطويل يتبيّن اختلاف ميولها.

القول على الارتفاع

[١٦٠] الارتفاع هو قوس من الدائرة السمتية* فيما بين النقطة
,تفعة وبين الأفق. [١٦١] فكلّ كوكب من الكواكب إذا ارتفع عن الأفق
ان مقدار ارتفاعه هو القوس من الدائرة العظيمة التي تخرج من سمت
رأس وتمرّ بالنقطة المسامتة لمركز الكوكب وتنتهي* إلى الأفق وأعني
لنقطة المسامتة النقطة التي هي طرف الخطّ الذي يخرج من مركز العالم إلى
ركز الكوكب وينتهي على استقامة إلى سطح الفلك الأعلى. [١٦٢] وكلّما
,داد الكوكب ارتفاعاً عظمت تلك القوس ولا يزال كذلك إلى أن ينتهي إلى
ائرة نصف النهار فتكون* القوس من دائرة نصف النهار وهي أحد الدوائر
سمتية أيضاً فيما بين النقطة المسامتة (٢١١/ك) لمركز (ب١٠٧/ي)
كوكب وبين الأفق هي نهاية ارتفاعه. [١٦٣] ثمّ ينحدر إلى ناحية المغرب
ينقص ارتفاعه لأنّ تلك القوس تصغر* ولا يزال كذلك وكلّما* قرب من
غرب نقص ارتفاعه إلى أن ينتهي إلى نقطة المغرب هكذا حال الشمس
جميع الكواكب الطالعة* والغاربة.

[١٦٤] فأما الكواكب الدائمة الظهور فإنّها تتحرّك* على دوائر ظاهرة
تكون* أبداً ظاهرةً ولكن يختلف ارتفاعها أيضا لأنّها تتحرّك* صاعدةً فلا
زال ارتفاعها يتزايد إلى أن تنتهي* إلى دائرة نصف النهار فعندها يكون
عظم ارتفاعها ثمّ تتحرّك منحدرةً فتنتهي* أيضاً إلى دائرة نصف النهار*
أنّ دائرة نصف النهار تقطع* الدائرة العظيمة التي يتحرّك* عليها
لكوكب على نقطتين ظاهرتين فإذا انتهي الكوكب في انحداره إلى دائرة
صف النهار فعندها حينئذ أقلّ ما يكون ارتفاعاً هذا هو الذي يسمّي
لارتفاع.

[١٦٥] والارتفاع على الحقيقة هو العمود الواقع من مركز (ب٢١/ك)
لكوكب على سطح الأفق هذا العمود هو جيب* القوس الموازية لقوس

الارتفاع الشبيهة بها التي تمرّ بمركز الكوكب وتنتهي* إلى الأفق. [١٦٦] وجيب كلّ قوس هو نصف وتر ضعف تلك القوس وهذا حدّ الجيب* المستوي. [١٦٧] فأما الجيب المعكوس* فهو سهم ضعف القوس وقد تستعمل* الجيوب في مواضع كثيرة لأنّ منها تستخرج* القسي. فأعظم ارتفاعات الكواكب هو نصف نهارها.

القول على الطالع والمطالع

[١٦٨] الطالع هو كلّ جزء يكون من دائرة البروج على محيط الأفق من ناحية المشرق والمغارب هو الجزء المقابل له من ناحية المغرب لأنّ دائرة البروج يقطعها الأفق بنصفين ويكون أبدأ على محيط الأفق نقطتان متقاطرتان من دائرة البروج. [١٦٩] فلذلك إذا* طلع جزء من دائرة البروج غرب الجزء المقابل له وكذلك كلّ دائرة عظيمة من الدوائر التي في الفلك.

[١٧٠] فأما الدوائر التي تتحرّك (٢٢أ/ك) عليها الكواكب الثابتة وسائر الدوائر المتوازية* فليس حالها كذلك أعني أنه ليس الجزء منها الذي في* المشرق مقابلاً للجزء الذي في المغرب في كلّ الآفاق فانّ الآفاق المائلة ليست* تقسم كلّ هذه الدوائر بنصفين بل بأقسام مختلفة الاّ ما كان منها دائرة عظيمة.

[١٧١] فأما* المطالع فهو* القوس التي تطلع* من دائرة معدّل النهار مع القوس المفروضة من دائرة البروج. [١٧٢] ونمثّل* في ذلك مثالاً وهو أن نفرض النقطة المسامتة لمركز الشمس على محيط الأفق فعند ذلك تكون* نقطة ما من* معدّل النهار على محيط الأفق ثمّ نتوهّم الشمس متحرّكةً في الارتفاع فتتحرّك* النقطة من معدّل النهار صاعدةً فإذا انتهت الشمس إلى ساعة ما* مفروضة من الدائرة التي تتحرّك* عليها تكون قد طلعت* في ذلك المقدار من الزمان قوس من دائرة البروج وهي التي من النقطة المسامتة لمركز الشمس إلى أفق المشرق فتكون* حينئذ النقطة من معدّل (ب٢٢/ك) النهار قد انتهت إلى جزء ما من معدّل النهار فتكون* القوس من معدّل النهار التي بين ذلك الجزء وبين الأفق هي مطالع تلك القوس* من دائرة* البروج.

[١٧٣] وهذه القوس من معدّل النهار شبيهة بالقوس من الدائرة*

الموازية لمعدّل النهار التي تحرّك على مسامتتها* مركز الشمس فتسمّي هذه القوس أيضاً مطالع تلك القوس من دائرة* البروج. وأريد بالشبيهة أن عدد أجزائها يكون متساوية*.

[١٧٤] فلكلّ* جزء من دائرة البروج جزء من معدّل* النهار هو مطالع ذلك الجزء ومطالعات* أجزاء دائرة البروج مختلفة إلاّ أنّ كلّ جزئين منها متساويين يكون بعدهما من أحد نقطتي الاعتدال بعداً واحداً (ب١٠٧/ي) فإنّ مطالعهما متساوية مثال ذلك العشر درجات* الأول من الحمل والعشر درجات* الأواخر من الحوت مطالعهما متساوية. [١٧٥] وكذلك جميع الحمل وجميع الحوت وكذلك أيضاً* السنبلة والميزان فإنّ* مطالعهما متساوية وكذلك جميع أجزاء البروج (٢٢١/ك) كلّ ربع منها تساوي مطالعه* مطالع الربع المتّصل به وكلّ جزء من أجزاء الربع تساوي مطالعه* مطالع الجزء النظير* له من الربع المتّصل به.

[١٧٦] فهذه حال المطالع في كلّ واحد من الآفاق إلاّ أنّ مطالع كلّ جزء من دائرة البروج في أفق من الآفاق مخالفة* لمطالع ذلك الجزء في غير ذلك الأفق واختلاف المطالع بحسب اختلاف العروض. [١٧٧] والمثال في ذلك مطالع الحمل إذا كان في أفق ما أجزاء ما فإنّ مطالعه في الأفق الذي عرضه أكثر من ذلك العرض أجزاء ما أقلّ من تلك الأجزاء وكذلك الحوت وبالضدّ من ذلك السنبلة والميزان. ومطالع النصف من دائرة البروج الذي بين نقطتي الاعتدال أبداً نصف دائرة معدّل* النهار.

[١٧٨] فأما في الأفق الذي عرضه ستّة وستّون جزءاً وهو الذي أطول* نهاره أربعة وعشرون ساعة ومدار السرطان فيه مماس للأفق* فإنّ في هذا* الأفق تنطبق* دائرة البروج على دائرة الأفق في كلّ يوم (ب٢٢/ك) وليله*. [١٧٩] ثمّ إذا بدأت بالحركة بعد الانطباق تطلع في آن* واحد ستّة بروج فلا مطالع لهذا النصف في ذلك الأفق ثمّ يبدأ الفلك بالحركة فيعود حال المطالع في ذلك الأفق كحالها في سائر الآفاق.

[١٨٠] وأما المكان الذي تحت القطب فليس يطلع فيه شيء من دائرة البروج ولا يغرب بل النصف منها أبداً ظاهر والنصف أبداً خفي. [١٨١] وأما المواضع التي بين الموضع الذي تحت القطب وبين الأفق الذي

عرضه ستّة وستّون درجة فإنّ بعض دائرة البروج يكون فيها أبداً ظاهراً* ولا مطالع له في تلك المواضع ونظيره من دائرة* البروج يكون أبداً خفياً ولا مطالع له في تلك المواضع . [١٨٢] وأما الذي بين الجزء الظاهر وبين الخزء الخفي* فطالع وغارب ومطالعه كمطالع غيره في سائر الآفاق كلّ قوس منها تطلع مع قوس من معدّل النهار وهذا الذي يطلع ويغرب في هذه (٢٤١/ك) الآفاق من دائرة البروج هو قوسان غير متّصلتين* يفصّل بينهما من أحد الطرفين الجزء الظاهر ومن الطرف الآخر الجزء الخفي واحدي هاتين القوسين تلي نقطة الاعتدال الربيعي وتفصّلها هذه النقطة بنصفين والأخرى تلي نقطة الاعتدال الخريفي تفصّلها هذه النقطة بنصفين. [١٨٣] والقوس التي تلي نقطة الاعتدال الربيعي يكون طلوعها في هذه الآفاق منكوساً وغروبها مستوياً والقوس التي تلي الاعتدال الخريفي يكون طلوعها مستوياً وغروبها منكوساً فهذه جملة القول على المطالع .

[١٨٤] وإذ قد أتينا على شرح هذه الدوائر المتوهّمة في سطح الفلك الأعلى فإنّا نخطّ لذلك صورةً نرسم فيها جميع الدوائر والخطوط التي ذكرناها ليكون الناظر فيها متصوّراً لجميع ما في سطح الفلك وهذه هي الصورة.

[١٨٥]* قد تقدّم القول في أن لكلّ كوكب من الكواكب بالجملة فلكاً* يخصّه* هو جزء من جميع الجسم المحيط وكلّ فلك من أفلاك الكواكب السيارة ينقسم إلى أجزاء فلنذكر حال كلّ كوكب منها وأفلاكه فلنبدأ بأشرفها* وهو فلك الشمس ثمّ نتبعه* بالذي يليه واحداً فواحداً.

القول على فلك الشمس

[١٨٦] فلك الشمس جسم مستدير يحدّه سطحان كريان متوازيان مركزهما مركز العالم السطح الأعلى منهما سطح محدّب مماس لمقعّر فلك المريخ والسطح الأدنى منهما (١٠٨١/سي) سطح مقعّر مماس لمحدّب فلك الزهرة. [١٨٧] وينفصل منه كرة مستديرة يحدّها أيضاً سطحان كريان متوازيان مركزهما نقطة واحدة خارجة عن مركز العالم والسطح الأعلى من هذه الكرة مماس للسطح الأعلى من الكرة الأولى على نقطة واحدة مشتركة لهما والسطح* الأدنى منها* مماس للسطح الأدنى أيضاً من الكرة* الأولى على نقطة واحدة مشتركة لهما ويسمّي هذا الفلك الفلك* الخارج المركز وهو يتحرّك* حركةً مستديرةً من جهة المغرب إلى جهة المشرق على قطبين ثابتين ومحور ثابت غير قطبي العالم ومحوره وغير قطبي دائرة البروج ومحورها*.

[١٨٨] والشمس جسم مصمت شكله شكل كرة مركوزة في جسم هذا الفلك بعد مركزها من قطبيه بعد واحد وسطحها الظاهر مماس لسطحيه على نقطتين أعني أن نقطتين من سطح جرم الشمس وهما طرفا قطرها مشتركتين* لسطحها ولسطحي فلكها الثاني فقطر كرة الشمس مساو لسمك جسم فلكها. وإذا تحرّك هذا الفلك من جهة المغرب إلى جهة المشرق على قطبيه* الثابتين حرّك معه جرم الشمس لأنّها كالجزء منه.

[١٨٩] فإذا تحرّك الشمس الحركة المستديرة رسمت نقطة مركزها دائرة متوهّمة مركزها مركز هذا الفلك فتسمّي هذه الدائرة أيضاً الفلك الخارج المركز. [١٩٠] فإذا توهّمنا سطح هذه الدائرة قاطعاً للعالم أحدث في سطح الفلك الأعلى دائرة هي دائرة البروج كما بيّنا آنفاً وأحدث في سطحي فلك الشمس الذي مركزه مركز العالم أيضاً دائرتين متوازيتين مركزهما مركز العالم تسمّي* العظمى منهما* الفلك الممثّل بفلك البروج.

[١٩١] وأحدث في سطحي فلك الشمس الخارج المركز دائرتين متوازيتين وموازيتين للفلك الخارج المركز تكون* العظمى منهما مماسة

للفلك الممثّل بفلك البروج وتكون* الصغرى مماسة للدائرة الموازية له.
[١٩٢] وأحدث في كلّ فلك من أفلاك الكواكب التي بها قسمنا أوّلاً جميع الفلك بتسعة أقسام متامسة مركزها مركز العالم دائرتين متوازيتين مركزهما مركز العالم تسمّي* العظمى منهما الفلك الممثّل بفلك البروج لذلك الكوكب وتكون الصغرى منهما مطابقة للعظمى من الفلك الذي يليه.
[١٩٣] ونقطتا التماس من الفلك الخارج المركز هما البعد الأبعد والبعد الأقرب من الفلك الخارج المركز أعني أن أحديهما* هي أبعد نقطة على محيط الفلك الخارج المركز والأخرى هي أقرب نقطة على محيطه من مركز العالم لأنّه إذا وصل بين مركز العالم وبين مركز فلك الشمس خطّ مستقيم وأُخرج على استقامته في الجهتين إلى نقطتي التماس* ويكون القسم الذي يلي المركز الخارج أطول الخطوط التي تخرج* من مركز العالم إلى محيط* الفلك الخارج المركز والقسم الباقي أقصر الخطوط التي تخرج* من مركز العالم إلى محيطه. [١٩٤] وهذا الخطّ يقطع محيط* الدائرة الخارجة المركز التي تمرّ بمركز الشمس التي تسمّي* أيضاً الفلك الخارج المركز على نقطتين متقابلتين فتسمّي التي على الخطّ الأطول البعد الأبعد أيضاً والتي على الخطّ الأقصر البعد الأقرب وتسمّي* نقطة البعد الأبعد أوج الشمس وتسمّي* نقطة البعد الأقرب حضيض الشمس.
[١٩٥] ونقطة الأوج مسامتة لنقطة من فلك البروج في جهة الشمال بعدها من نقطة الانقلاب الصيفي أربعة وعشرون جزءاً ونصف بالأجزاء التي بها الدائرة ثلاثمئة وستّون جزءاً متقدّمة عنها بهذه الأجزاء على ما ذكره بطلميوس في كتابه* المرسوم بالمجسطي. [١٩٦] وهذه* النقطة على ما ذكر ثابتة غير منتقلة* عن مسامتة النقطة من فلك البروج (ب١٠٨/ي) التي بعدها من الانقلاب* الصيفي هذه الأجزاء. [١٩٧] ونقطة الحضيض في جهة الجنوب متقدّمة عن نقطة الانقلاب الشتوي بمثل هذه الأجزاء. [١٩٨] فأمّا على ما ذكره المتأخّرون من أصحاب الرصد فإنّهما يتحرّكان وينتقلان على توالي البروج.
[١٩٩] وإذا كانت الشمس تتحرّك* بحركة فلكها حركة مستديرة فإنّ مركزها لا يزال يسامت* جزءاً بعد جزء من أجزاء دائرة البروج أعني أن

الخطوط التي تخرج* من مركز العالم إلى* مركز الشمس وتنتهي* إلى دائرة البروج وقتاً بعد وقت تنتهي* إلى نقطة بعد نقطة فلا يزال كذلك إلى أن يعود إلى مسامتة النقطة الأولى فعنده تكامل سنة الشمس.

[٢٠٠] وإذا كانت الشمس أبداً لازمةً لمحيط الفلك الخارج المركز فإنّها إذا كانت في البعد الأبعد من فلكها أعني الأوج فهي أبعد ما تكون* من الأرض فاذا انتقلت فلا تزال تقرب* إلى أن تنتهي* إلى حضيض فلكها فحينئذ هي أقرب ما تكون* من الأرض ثمّ لا تزال* صاعدة حتى تعود* إلى أوجها كذلك دائماً فلهذه الحال صارت أبعادها من الأرض مختلفة.

[٢٠١] فصارت تسامت من دائرة البروج قيساً متساويةً في أزمنة مختلفة أعني أنّها تقطع أجزاء النصف من دائرة البروج الذي يلي الأوج في زمان أكثر من الزمان الذي تقطع* فيه أجزاء النصف الذي يلي الحضيض.

[٢٠٢] وكذلك أيضاً صار للشمس حركتان أحديهما حركتها الذاتية وهي حركة فلكها الخارج المركز حول مركزه وعلى قطبيه وهي الحركة المستوية والأخرى حركتها التي توجد بالاضافة إلى دائرة البروج التي مركزها مركز العالم وهي الحركة المختلفة.

[٢٠٣] ولأنّ الشمس إذا كانت في حضيضها تكون* أقرب ما تكون* من الأرض وجب أن يكون الموضع من الأرض المسامت لحضيضها أشدّ الأماكن حرارةً وهذا المكان من الأرض هو في جهة الجنوب قريب من الخطّ المسّمي خطّ الاستواء مسامت للنقطة التي قدّمنا ذكرها* وهو مستدير حول الأرض. [٢٠٤] لأنّ الشمس في الزمان الذي تكون* في حضيضها* يحرّكها الفلك الأعلى الحركة السريعة فتدور في اليوم والليلة حول الأرض على دائرة موازية لمعدّل النهار وهي في أقرب قربها فتسامت من الأرض أيضاً دائرة هي أقرب الدوائر إلى تلك الدائرة.

[٢٠٥] ولذلك أنا إذا توهّمنا خطّاً مستقيماً يخرج من مركز العالم إلى مركز الشمس وهي في حضيضها فإنّه يلقي سطح الأرض على نقطة. فإذا تحرّكت الشمس حول الأرض وتوهّمنا ذلك الخطّ متحرّكاً معها فإنّ النقطة منه التي على سطح الأرض أيضاً تحدث* دائرة على سطح الأرض وليست* دائرة عظيمة لأنّ الدائرة الزمانية التي تتحرّك عليها الشمس وهي في

حضيضها ليست دائرة عظيمة. [٢٠٦] والخطّ المستقيم الذي يخرج إليها من مركز العالم المتحرّك معها ليس* يتحرّك في سطح مستوي بل على سطح مخروط فلذلك تكون* القطعة* التي تدور حولها من مركز الأرض ليست نصف كرة فتكون* الدائرة التي تحدث من حركته على سطح الأرض ليست دائرة عظيمة. [٢٠٧] ولأنّ ذلك الخطّ أعني الخطّ الذي يخرج من مركز العالم إلى حضيض الشمس أقصر الخطوط التي تخرج إلى الشمس ونصف قطر الأرض ليس* يختلف تكون القطعة منه التي بين سطح الأرض وبين الحضيض أقصر الخطوط التي توجد في طول السنة بين مركز الشمس وبين سطح الأرض فلذلك تكون* هذه الدائرة التي ذكرنا أقرب الدوائر من الشمس ومن أجل ذلك تكون* أشدّ أماكن الأرض حرارةً.

[٢٠٨] وكلّما قرب منها كان أيضاً شديد الحرارة وتسمّي* هذه الدائرة الطريقة المحترقة ولذلك صارت الجهة الجنوبية أشدّ حرارةً من الجهة الشمالية وصار (١٠٩١/ي) أيضاً خطّ الاستواء* من الأرض شديد الحرارة لقربه* من هذه الطريقة ويقال إنّ هذه الأماكن من الأرض لشدّة حرارتها غير مسكونة ولا فيها شيء من الحيوان.

[٢٠٩] ولنخطّ* صورةً لفلك الشمس على الدائرة التي تحدث* بحركة الشمس ولجميع الدوائر التي تحدثها* وهي هذه.

القول على فلك القمر

[٢١٠] فلك القمر جسم كريّ مستدير يحدّه سطحان كريّان مركزهما مركز العالم السطح الأعلى منهما مماس لمقعّر فلك عطارد والسطح الأدنى منهما مماس لكرة النار. [٢١١] وفي السطح* الأعلى منهما* يكون الفلك المثّل بفلك البروج الذي قدّمنا ذكرها. [٢١٢] وهذا الفلك أعني فلك القمر يتحرّك بجملته حركة مستديرة من جهة المشرق إلى جهة المغرب على خلاف توالي البروج على قطبين ثابتين مماستين لقطبي دائرة البروج.

[٢١٣] وينقسم هذا الفلك أوّلاً قسمين كريين مستديرين متماسين مركزهما مركز العالم يحيط أحدهما بالآخر ويماس السطح المقعّر من القسم الأعلى بالسطح المحدّب من القسم الأدنى. [٢١٤] والقسم الأدنى وهو الأقرب إلى مركز العالم يتحرّك حركة مستديرة من جهة المشرق من جهة المشرق أيضاً إلى جهة المغرب على خلاف توالي البروج على قطبين ثابتين غير قطبي فلك البروج ويسمّي هذا الفلك المائل.

[٢١٥] وينفصل من هذا الفلك كرة مستديرة يحدّها سطحان كريّان مستديران مركزهما خارج عن مركز العالم وعن سطح الفلك المثّل بفلك البروج وبعده عن قطبي الفلك المائل بعداً واحداً والسطح الأعلى منهما مماس للسطح الأعلى من الفلك المائل على نقطة واحدة والسطح الأدنى* منهما مماس للسطح الأدنى من الفلك المائل أيضاً على نقطة واحدة ويسمّي هذا الفلك الفلك الخارج المركز. [٢١٦] وهذا الفلك يتحرّك حركة مستديرة من جهة المغرب إلى جهة المشرق على توالي البروج على قطبين ثابتين غير الأقطاب التي تقدّمت .

[٢١٧] وينفصل من هذا الفلك كرة مستديرة بين سطحيه المتوازيين بعد مركزها من قطبيه بعد واحد وقطره مساو لسمك الفلك الخارج وسطحها الظاهر مماس لسطحيه المتوازيين على نقطتين مشتركتين* وهي مركوزة في جسم هذا الفلك كالجزء منه وهي أيضاً تتحرّك* حركة

مستديرة حول مركزها على قطبين ثابتين لها وهذه الكرة تسمّي* فلك تدوير. [٢١٨] ويسمّي الفلك الخارج المركز الفلك الحامل لأنّه كالحامل لهذه الكرة.

[٢١٩] والقمر جسم مصمت شكله شكل كرة مركوزة في جسم فلك التدوير سطحه الظاهر مماس لسطح فلك التدوير على نقطة واحدة. [٢٢٠] فإذا تحرّك الفلك الخارج المركز الحامل لفلك التدوير على قطبيه تحرّك* بحركته جميع فلك التدوير بكليته لأنّه كالجزء منه وانتقل معه جرم القمر لأنّه متّصل* به.

[٢٢١] وإذا تحرّك الكرة حركة مستديرة فكلّ نقطة فيها ترسم دائرة متوهّمة فمركز فلك التدوير يرسم* بحركة الفلك الخارج المركز دائرة متوهّمة مركزها مركز الخارج المركز تسمّي هذه أيضاً الفلك الخارج المركز وتسمّي الفلك الحامل لأنّ مركز فلك التدوير يتحرّك على محيطها.

[٢٢٢] فإذا توهّمنا سطح هذه الدائرة قاطعاً للعالم فإنّه يحدث في سطح الفلك الخارج المركز دائرتين متوازيتين وموازيتين للدائرة الأولى يعني الوسطى ومركزهما مركز الفلك الخارج المركز ويحدث في سطح فلك التدوير أيضاً دائرة مماسة للدائرتين المتوازيتين على نقطتين* مركزها وهو مركز فلك التدوير على محيط الدائرة الأولى . [٢٢٣] والدائرة الموازية لهذه الدائرة التي تمرّ بمركز القمر تسمّي* أيضاً فلك التدوير وعليها (ب١٠٩/ي) يتحرّك مركز القمر.

[٢٢٤] ويحدث أيضاً في سطح الفلك المائل دائرتين مركزهما مركز العالم وقطباهما قطبا الفلك المائل ويحدث أيضاً في سطح الفلك الأوّل المحيط بالفلك المائل أيضاً دائرة موازية للدائرتين الأوّليتين ومقاطعة للفلك الممثّل بفلك البروج. [٢٢٥] ويحدث أيضاً في سطح الفلك الأعلى دائرة مقاطعة لدائرة البروج تسمّي هذه الدائرة أيضاً الفلك المائل للقمر. [٢٢٦] وميل هذه الدائرة عن دائرة البروج هو الذي يسمّي عرض القمر لأنّ مركز القمر يكون أبداً في سطح هذه الدائرة.

[٢٢٧] فأما نقطتا التقاطع لهذه الدائرة ولدائرة البروج وهما اللتان تسمّيان* الجوزهران أما التي هي مبدأ النصف الشمالي من* الدائرة المائلة

فتسمّي* الرأس وأما التي هي مبدأ النصف الجنوبي فتسمّي* الذنب وميل هذه الدائرة عن دائرة البروج ميل واحد أبداً غير متغير.

[٢٢٨] وإذا أخرج خطّ مستقيم من مركز العالم إلى مركز الفلك الحامل وأخرج على استقامته من الجهتين إلى محيط الدائرة التي هي الفلك الخارج المركز فالنقطة التي هي طرف الخطّ التي تلي* المركز الخارج وهي أبعد النقط من مركز العالم تسمّي أوج القمر والمقابلة لها تسمّي حضيض القمر. [٢٢٩] وإذا كان مركز التدوير على البعد الأبعد وأخرجنا الخطّ على استقامته إلى محيط فلك التدوير فإنّ نهايته تسمّي أيضاً البعد الأبعد لفلك التدوير والطرف الآخر من قطره يسمّي البعد الأقرب لفلك التدوير.

[٢٣٠] فالفلك الأوّل من أفلاك القمر يتحرّك حركة مستديرة على قطبيه المسامتين لقطبي دائرة البروج وهما قطبا الفلك الممثّل بفلك البروج فيحرّك* بحركته جميع الأفلاك التي فيه* فينتقل الفلك بحركته على خلاف توالي البروج وينتقل جميع سطح الدائرة المائلة على خلاف توالي البروج وهذه الدائرة هي التي قلنا إنّ محيطها يقاطع محيط دائرة البروج. [٢٣١] فينتقل بهذه الحركة نقطتا التقاطع على خلاف توالي البروج وهذه الحركة هي التي تسمّي حركة الجوزهر.

[٢٣٢] والفلك الثاني من أفلاك القمر وهو الفلك المائل يتحرّك أيضاً حركة مستديرة على خلاف توالي البروج على قطبيه الثابتين وهما قطبا الدائرة المائلة التي في سطحه فيحرّك* بحركته جميع ما فيه فينتقل الفلك الخارج المركز بهذه الحركة على خلاف توالي البروج وينتقل جميع سطح الدائرة التي هي الفلك الخارج المركز بانتقاله وينتقل مركزها وينتقل البعد الأبعد وتكون* حركة هذه كلّها على خلاف توالي البروج حول مركز العالم وفي سطح الفلك المائل.

[٢٣٣] والفلك الخارج المركز وهو الفلك الحامل يتحرّك أيضاً حركة مستديرة على توالي البروج على قطبيه الثابتين فينتقل بحركته فلك التدوير على توالي البروج وينتقل مركزه على محيط الفلك الخارج المركز. [٢٣٤] وينتقل بانتقال فلك التدوير جرم القمر على توالي البروج .

[٢٣٥] وفلك التدوير يتحرّك أيضاً حركة مستديرة على قطبين له

ثابتين فيدور حول مركزه ويتحرّك بحركته جرم القمر ويتحرّك مركز القمر على محيط الدائرة التي تسمّي أيضاً فلك التدوير. [٢٢٦] أما إذا تحرّك أعني مركز القمر على القوس الذي يلي البعد الأبعد من فلك التدوير فإنّ حركته تكون* على خلاف توالي البروج وأما إذا تحرّك على القوس الذي يلي البعد الأقرب فإنّ حركته تكون* على توالي البروج .

[٢٢٧] فأما قطر فلك التدوير الذي نهايته البعد الأبعد الذي كان على استقامة قطر الفلك الخارج المركز فإنّه إذا تحرّك فلك التدوير وانتقل* مركزه عن* نقطة البعد الأبعد من الفلك الخارج المركز وانتقل معه قطره الذي طرفاه البعد الأبعد والبعد الأقرب فليس يبقي هذا القطر على وضعه الذي كان له أعني أنّه ليس (١١٠١/ي) يكون أبداً مسامتاً لمركز الفلك الخارج المركز ولا مسامتاً لمركز العالم. [٢٢٨] أريد أنّه إذا خُرج على استقامته لم ينته إلى مركز العالم ولا إلى مركز الفلك الخارج المركز ولكنّه أبداً مسامتاً لنقطة اخرى من ذلك القطر مما يلي القرب الأقرب من الفلك الخارج المركز يكون بعدها من مركز العالم بمقدار مساو لبعد مركز الفلك الخارج المركز منه وحافظاً لهذه المسامتة أبداً.

[٢٢٩] وأما مركز فلك التدوير ونقطة البعد الأبعد من الفلك الخارج المركز فإنّ حركتيهما اللتين ذكرناهما متضاتين في جهتين مختلفتين* أحديهما إلى توالي البروج فهي حركة مركز فلك التدوير والأخرى إلى خلاف توالي البروج وهي حركة البعد الأبعد إلاّ أنّ بعدهما أبداً من موضع الشمس الأوسط يكون بعداً متساوياً أعني أن موضع الشمس يكون متوسّطاً بين هاتين النقطتين.

[٢٤٠] ونقطة البعد الأبعد تتحرّك* حول مركز العالم فتنتقل* من مسامته موضع الشمس إلى أن تعود* إلى مسامته في زمان الشهر القمري. [٢٤١] ومركز فلك التدوير يتحرّك على محيط الفلك الخارج المركز فينتقل من نقطة البعد الأبعد ويعود إليها في زمان الشهر القمري دفعتين ففي وقت الاجتماع الوسط يكونان جميعاً متّفقين فيجتمع مركز فلك التدوير ونقطة البعد الأبعد ثم يتحرّكان في جهتين مختلفتين* .

[٢٤٢] وإذا قطع مركز فلك التدوير نصف محيط الفلك الخارج

المركز وصار في القرب الأقرب وصار مقابلاً* لنقطة البعد الأبعد تكون* نقطة البعد الأبعد قد تحرّكت من مسامتة موضع الشمس وصار بعدها عنها ربع دائرة فتكونان* النقطتان جميعاً على تربيع الشمس ولذلك يكون موضع القمر الوسط في التربيعات في قربه الأقرب من الفلك الخارج المركز. [٢٤٣] فإذا تحرّك مركز فلك التدوير فتمّم دائرته تحرّك البعد الأبعد أيضاً ربع دائرة فيتّفقان جميعاً في مقابلة الشمس فلذلك يكون موضع القمر الأوسط في الاستقابلات في بعده الأبعد من الفلك الخارج المركز. [٢٤٤] فموضع القمر أبداً في الاجتماع والاستقبال في بعده الأبعد من الفلك الخارج المركز وفي التربيعات في قربه الأقرب منه.

[٢٤٥] وأما جرم القمر فإنّه يتحرّك حول مركز فلك التدوير تارةً إلى خلاف توالي البروج وذلك إذا كان في أعلى فلك التدوير وتارةً إلى توالي البروج وذلك إذا كان فيما دنا من فلك التدوير إلاّ أنّه إذا كان متحرّكاً إلى خلاف توالي البروج فليس يُرى راجعاً لأنّ فلك التدوير يتحرّك على خلاف توالي البروج ويحرّك جرم القمر بحركته وحركته أسرع من حركة القمر المضادة له فليست تكون* حركة القمر لذلك متأخّرة أبداً ولكن قد يؤثر خلاف* حركته فتكون* حركته إذا كان في أعلى فلك التدوير أبطأ وتكون* إذا كان فيما دنا من فلك التدوير أسرع.

[٢٤٦] فالحركات التي تخصّ القمر ستّ حركات: حركة جميع فلكه إلى خلاف توالي البروج وهي التي تسمّي حركة الجوزهر؛ وحركة الفلك المائل إلى خلاف توالي البروج أيضاً على قطبين أخرين وتسمّي حركة الأوج؛ وحركة الفلك الخارج المركز المحرّك لفلك التدوير على توالي البروج مسامتاً لمركز فلك التدوير وعلى* محيط الفلك المائل وتسمّي* حركة العرض وهذه الحركة بعينها إذا أُخذت بالاضافة إلى مسامتة دائرة البروج عرضت الحركة التي تسمّي حركة الطول؛ وحركة فلك التدوير المحرّك* لجرم القمر حول مركز فلك التدوير وتسمّي حركة الاختلاف؛ وحركة قطر فلك التدوير الذي نهايته البعد الأبعد إلى مسامتة النقطة المفروضة وتسمّي* هذه الحركة حركة المحاذة.

[٢٤٧] فلنتوهّم* خطّاً مستقيماً يخرج (ب١١٠/ي) من مركز العالم

إلى مركز القمر ويخرج على استقامة فهو ينتهي إلى محيط الفلك المائل لأنّ مركز القمر أبداً في سطح الفلك المائل ونتوهّم أيضاً دائرة عظيمة تخرج* من قطبي فلك البروج وتمرّ بالنقطة التي على محيط الفلك المائل التي ينتهي إليها الخطّ المستقيم. [٢٤٨] وهذه الدائرة تقطع* محيط دائرة البروج أيضاً فتحصل بين دائرة البروج وبين الفلك المائل قوس من هذه الدائرة فهذه القوس تسمّي عرض القمر والنقطة التي عليها تقاطعت هذه الدائرة ودائرة البروج هي موضع القمر من دائرة البروج.

[٢٤٩] فإذا تحرّك القمر بالحركات التي تقدّمت فانتقل بما يجتمع له من تلك الحركات على توالي البروج فانتقل* الخطّ المستقيم الذي توهّمناه مارّاً بمركزه بتلك الحركات أيضاً. [٢٥٠] وانتقلت الدائرة أيضاً التي تمرّ بقطبي دائرة البروج على توالي البروج فتتحرّك* نقطتها* على محيط دائرة البروج فتكون* من ذلك الحركة التي تُرى للقمر في الطول بالاضافة إلى دائرة البروج. [٢٥١] وانتقال الخطّ المستقيم المارّ بمركز القمر بالاضافة إلى محيط الفلك المائل هو الحركة التي تُرى للقمر في العرض وبهذه الحركة تتغيّر* القوس التي بين الدائرتين التي هي العرض فتارةً تزيد وتارةً أخرى تنقص* فلذلك يختلف عرض القمر.

[٢٥٢] وإذا انتهت نقطة التقاطع لدائرة البروج وللدائرة* المارّة بقطبيها المتحرّكة على محيط دائرة البروج التي هي موضع القمر في الطول إلى النقطة المسامتة لمركز الشمس فهي الحال التي تسمّي الاجتماع وإذا انتهت إلى النقطة المقابلة لمركز الشمس فهي الحال التي تسمّي الاستقبال.

[٢٥٣] وإذا انتهت الخطّ المستقيم المارّ بمركز القمر المتحرّك في سطح الفلك المائل إلى نقطة التقاطع بين الفلك المائل وبين دائرة البروج وهي التي تسمّي الجوزهر واتّفق أن يكون مركز الشمس مسامتاً لتلك النقطة فيصير مركز الشمس ومركز القمر على ذلك الخطّ المستقيم فعند هذه الحالة يكون كسوف الشمس. [٢٥٤] وذلك أنّ الشعاع الذي يخرج من أبصارنا إلى جرم القمر يكون على ذلك الخطّ المستقيم أو قريب منه فيمنعه جرم القمر أن يمتدّ على استقامة حتّى ينتهي إلى جرم الشمس فيصير جرم القمر حاجزاً بين أبصارنا وبين جرم الشمس فيستر عنّا ضوء الشمس.

[٢٥٥] والشعاع يخرج من أبصارنا على شكل مخروط رأسه نقطة البصر وقاعدته سطح جرم المبصر فإذا نظرنا في وقت الكسوف إلى جرم القمر كان الشعاع الذي يخرج من أبصارنا مخروطاً رأسه نقطة أبصارنا وقاعدته جرم القمر . [٢٥٦] ونتوهّم سطح ذلك المخروط يخرج على استقامة ويمتدّ إلى ناحية الشمس فإن وقع جرم الشمس في داخل ذلك المخروط استترت الشمس كلّها عنّا فيقال لذلك إنّ جميع الشمس قد انكسف فإن كان الذي يقع في داخل المخروط بعض جرم الشمس ويكون البعض خارجاً عنه* استتر عنّا ذلك البعض وقيل إنّ الذي انكسف من الشمس هو ذلك البعض الذي في داخل المخروط.

[٢٥٧] وإذا انكسف جميع الشمس فليس يلبث حتّى يبتدئ بالاسفار* لأنّ مخروط الشعاع الذي ذكرناه إذا انتهي إلى فلك الشمس فإنّ قطره يكون مساوياً لقطر جرم الشمس فحين يبتدئ القمر بالحركة بعد الامتلأ ويتحرّك المخروط فتبتدئ* الشمس بالاسفار*.

[٢٥٨] وإذا انتهي الخطّ المستقيم إلى نقطة التقاطع واتّفق أن يكون مركز الشمس مسامتاً للنقطة الأخرى المقابلة لها فيصير* مركز الشمس ومركز القمر ومركز الأرض على ذلك الخطّ المستقيم فعند ذلك الحال يكون كسوف القمر. [٢٥٩] وذلك أنّ القمر لا نور له وإنّما (١١١١/ي) يكتسب النور من ضوء الشمس وهو جسم صقيل إذا قابلته الشمس قبل نورها واستنار* بضوءها وانعكس ذلك النور من سطحه إلى الأرض فانارت به.

[٢٦٠] والأرض جسم كثيف له ظلّ وظلّها أبداً يكون مقابلاً لموضع الشمس وشكل هذا الظلّ أيضاً شكل مخروط وذلك أنّ قطر الشمس أعظم من قطر الأرض والشعاعات التي تخرج من محيط جرم الشمس إلى محيط جرم الأرض تلتقي* بشكل مخروط يكون سطح الشعاع محيطاً به ويكون السطح المخروط من لدن جرم الأرض إلى نقطة الالتقاء ظلاً متّصلاً مظلّماً لا ضوء فيه لأنّ جرم الأرض قد ستر ضوء الشمس عنه وهذا الظلّ هو الذي يسمّي الليل. [٢٦١] وهذا المخروط يتجاوز فلك القمر فإذا صار جرم الشمس وجرم الأرض وجرم القمر على خطّ واحد مستقيم وكانت الأرض متوسّطة بينهما صار جرم القمر في مخروط الظلّ لما قدّمنا أن الظلّ أبداً

مقابل لجرم الشمس وهو يتجاوز فلك القمر فإذا صار في مخروط الظلّ صارت الأرض حاجزةً بينه وبين الشمس فتمنعه عن قبول النور ويعود* إلى جوهره فيُرى حينئذ مظلّما لا ضوء له.

[٢٦٢] وإذا كانت الأجرام الثلاثة على حقيقة الخطّ المستقيم فإنّ القمر يكون في وسط ذلك المخروط وإن كان القمر قريباً من ذلك الخطّ الذي عليه الجرمان الباقيان فإنّه يكون في جانب ذلك المخروط وعلى حسب قربه من ذلك الخطّ يكون مقدار ما يظلم منه ومقدار مكثه. [٢٦٣] فإذا كان جميع القمر في مخروط الظلّ صار جميعه مظلماً وقيل قد انكسف جميعه فإن كان بعضه في داخل المخروط وبعضه خارجاً عنه كان بعضه مظلماً وبعضه مستنيراً وقيل قد انكسف بعضه. [٢٦٤] فإذا انكسف جميعه وكان مارّاً بوسط المخروط طال مكثه وإن كان مارّاً في جانب المخروط قلّ مكثه لأنّ خروجه منه يكون أسرع.

[٢٦٥] فأما أن* يكون في بعض الأوقات جميع القمر مستنيراً وفي بعضها يكون بعضه* مستنيراً وفي بعضها لا يُرى منه شيء مستنيراً فإن ذلك بحسب أوضاعه من الشمس ومنّا فإن الذي يضيء من القمر قطعة قريبة* من نصف والذي يدركه أبصارنا هو أيضاً قريب من نصف إلاّ أنّ النصف المضي هو أبداً النصف المقابل للشمس المواجه لها لكن هذا النصف مختلف أوضاعه عندنا . [٢٦٦] أما في الاستقبالات فيكون النصف المواجه للشمس هو النصف الذي يلينا لأنّه في الاستقبالات مقاطر للشمس ونحن في هذا الموضع المتوسّط بينهما فنرى* من القمر جزءاً جميعه مضيء. [٢٦٧] فأما إذا انتقل عن المقابلة فلا يكون مكاننا* متوسّطاً بينهما بل يكون زائلاً عن الوسط بالاضافة إليهما فيصير النصف المضيء ليس هو النصف بعينها المواجه لنا بل يكون النصف المواجه لنا بعضه من الجزء المضيء وبعضه من الجزء المظلم. [٢٦٨] فلا يزال يتناقص المقدار المقابل لنا إلى أن ينتهي إلى الاجتماع فعند ذلك يكون النصف المضيء هو النصف الأعلى من القمر وهو الذي يلي الشمس ويكون النصف الذي يلينا جميعه مظلماً. [٢٦٩] فإذا فارق الشمس عاد إلى الحال الأوّل وصار النصف المضيء مليئاً* فتصير الدائرتان أعني الدائرة التي تحيط* بالجزء* المضيء والدائرة التي تحيط

بالجزء المظلم الذي يلينا متقاطعتين فيظهر لنا أوّلاً جزء يسير. [٢٧٠] وهذا الجزء اليسير يحيط به قوسان أحديهما القوس من الدائرة التي تحدّ الجزء الذي يلينا والأخرى القوس من الدائرة التي تحدّ الجزء المضيء لأنّ هذا الجزء اليسير يحيط به قوسان وهو على سطح كرة يكون هلالاً فلذلك صار الذي (ب١١٢/ي) يُرى من القمر في أوّل الشهر وآخره* هلالياً.

[٢٧١] ثمّ لا يزال يتزايد فإذا صار الضوء إلى النصف من الجزء الذي يلينا صارت حدبة القوس التي تحدّ الجزء المضيء مستقبلة لنا فيُرى القوس خطّاً مستقيماً ويصير ما يُرى من المقدار المضيء يحيط به خطّ مستقيم ونصف محيط الدائرة التي تحدّ الجزء الذي يلينا ولذلك صار المقدار المضيء الذي يُرى من القمر في هذه الأوقات نصف دائرة ثمّ لا يزال يتزايد إلى أن يعود إلى امتلائه.

[٢٧٢] فهذه* جملة القول على أفلاك القمر وأحواله ولكي* يكون ما ذكرناه واضحاً* فلنخطّ صورة لفلك القمر نرسم فيها جميع الدوائر التي ذكرناها وسائر الأحوال التي وصفناها.

القول على فلك عطارد

[٢٧٣] (١٢٥/ك) فلك عطارد جسم كريّ مستدير* يحدّه سطحان كريّان متوازيان مركزهما مركز العالم السطح الأعلى منهما مماس لمقعّر فلك الزهرة والسطح الأدنى منهما مماس لمحدّب فلك القمر. [٢٧٤] وفي السطح الأعلى منه يكون الفلك الممثّل بفلك البروج الذي حدث بحركة فلك الشمس.
[٢٧٥] وهذا الفلك أعني فلك عطارد يتحرّك حركة مستديرة حول مركز فلك البروج وعلى قطبيه من جهة المغرب إلى جهة المشرق وعلى توالي البروج حركة بطئية.
[٢٧٦] وتنفصل* من هذه الكرة كرة مستديرة يحدّها ويفرزها* سطحان كريّان متوازيان مركزهما وهو مركز الكرة جارج عن مركز العالم وعن سطح الفلك* الممثّل بفلك البروج وسطحها الأعلى مماس للسطح الأعلى لي مماس للسطح الأعلى من الفلك الأوّل على نقطة واحدة مشتركة* وهي تتحرّك على قطبين ثابتين غير قطبي فلك* البروج وحول مركزها الخارج حركة مستديرة إلى خلاف توالي البروج ويسمّي هذا الفلك الفلك المدير للفلك الحامل*.
[٢٧٧] وتنفصل* من هذا (ب٢٥/ك) الفلك كرة أخرى مستديرة يحدّها أيضاً سطحان كريّان متوازيان مركزهما وهو مركز الكرة جارج عن مركز العالم وعن مركز الفلك المدير أيضاً وعن سطح الفلك الممثّل بفلك البروج وفي الجهة أيضاً التي فيها مركز المدير وبعده من مركز المدير نصف البعد بين مركز المدير وبين مركز العالم وسطحها الأعلى مماس للسطح الأعلى من الفلك المدير على نقطة واحدة ويسمّي هذا الفلك الخارج المركز الفلك الحامل* وهذا الفلك يتحرّك أيضاً على قطبين ثابتين غير الأقطاب الأوّل حركة مستديرة على* توالي البروج*.
[٢٧٨] وتنفصل* من هذه الكرة كرة أخرى مستديرة* فيما بين سطحيه المتوازيين بعد مركزها من قطبيه الثابتين* بعد متساو وسطحها

الظاهر مماس للسطحين المتوازيين على نقطتين مشتركتين* وقطرها مساو للبعد الذي بين السطحين المتوازيين* وتسمّي* هذه الكرة فلك التدوير وهذه الكرة أيضاً تتحرّك على قطبين ثابتين لها حركة مستديرة حول مركزها. [٢٧٩] وجرم الكوكب أعني عطارد جسم مصمت شكله شكل كرة مركوز في جسم فلك التدوير سطحه الظاهر مماس لسطح فلك التدوير على نقطة واحدة.

[٢٨٠] فإذا (٢٦١/ك) تحرّك الفلك الحامل على قطبيه الخاصّين له حرّك بحركته جميع فلك التدوير ونقله بنقلته حول مركزه فحدث من حركة* مركزه دائرة متوهّمة مركزها مركز الفلك الحامل وليست في سطح الفلك المثّل بفلك البروج. [٢٨١] فإذا توهّمنا هذا السطح قاطعاً للعالم فإنّه يحدث في سطحي الفلك الحامل المتوازيتين دائرتين متوازيتين* وموازيتين للدائرة الأولى مركزها مركز الحامل.

[٢٨٢] ويحدث* أيضاً في سطحي* الفلك الأوّل الذي* مركزه مركز العالم دائرتين متوازيتين مركزهما مركز العالم العظمى منهما وهي التي في السطح الأعلى مقاطعة للفلك المثّل بفلك البروج على نقطتين وتسمّي* هذه الدائرة الفلك المائل.

[٢٨٣] ويحدث أيضاً في سطح الفلك الأعلى دائرة* (١١٢١/ي) مقاطعة* لدائرة البروج على نقطتين متقابلتين تسمّيان* الجوزهرين* وتسمّي* هذه الدائرة أيضاً الفلك المائل.

[٢٨٤] ولنتوهّم خطّاً مستقيماً يخرج من مركز العالم إلى الفلك المدير وينتهي إلى محيطه من الجهتين وليكن عليه أيضاً مركز الفلك الحامل (ب٢٦/ك) في الجهة من طرف الخطّ التي* فيها مركز المدير وطرف* هذا الخطّ الذي يلي الجهة التي عليها المركزان فهو* نقطة التماس من سطح الفلك المدير وهو البعد الأبعد من الفلك الخارج المركز وهو أبعد نقطة في فلك عطارد من مركز العالم ومقابله* البعد الأقرب منه.

[٢٨٥] فالفلك الأوّل من أفلاك عطارد المحيط بجميع أفلاكه يتحرّك حركة مستديرة حول مركز العالم على قطبين مسامتين لقطبي فلك البروج فيحرّك* جميع ما يحيط به من أفلاك هذا الكوكب وينقله وينقل مركزي

الفلك الحامل والمدير وينقل البعد الأبعد والبعد الأقرب وينقل نقطتي* التقاطع.

[٢٨٦] وهذه الحركة حركة بطيئة وهي على توالي البروج ومبلغها على ما ذكر بطلميوس في كتابه في التعليم في كلّ مائة سنة جزء واحد* من أجزاء الدائرة التي هي ثلاثمئة وستّون جزءاً.

[٢٨٧] والفلك الثاني من أفلاك عطارد وهو الفلك المدير يتحرّك حول مركزه على قطبين ثابتين* فيحرّك* جميع الفلك الحامل ويحرّك* (٢٧١/ك) مركزه وجميع ما فيه حركة مستديرة على خلاف توالي البروج حول مركز الفلك المدير. [٢٨٨] والفلك الحامل يتحرّك أيضاً حركة مستديرة حول مركزه وعلى قطبيه الخاصّين من جهة المغرب إلى جهة المشرق فيحرّك فلك التدوير بحركته. [٢٨٩] ويحرّك مركزه على محيط الدائرة التي تسمّي أيضاً الفلك الحامل حركة مستديرة على توالي البروج. [٢٩٠] ويتحرّك مع جملة فلك التدوير جرم الكوكب أيضاً على توالي البروج.

[٢٩١] وفلك التدوير أيضاً يتحرّك حول مركزه حركة مستديرة فيحرّك* جرم الكوكب أعني عطارد حركة مستديرة. [٢٩٢] وتحدث* من حركة مركزه دائرة متوهّمة مركزها مركز فلك التدوير تسمّي* أيضاً فلك التدوير وهذه الدائرة مائلة على سطح الفلك الخارج المركز. [٢٩٣] ومركز الكوكب يتحرّك أبداً على محيط هذه الدائرة أما إذا كان في أعلاها فإنّ حركته* تكون* إلى توالي البروج وأما إذا كان في أدناها* فإنّ حركته تكون* على خلاف توالي البروج.

[٢٩٤] وإذا تحرّك الفلك الحامل فحرّك (ب٢٧/ك) جميع فلك التدوير فإنّ قطره الذي كان على استقامة الخطّ الذي عليه مراكز الأفلاك ينتقل ويتغيّر* وضعه.

[٢٩٥] فنتوهّم* نقطة من منتصف الخطّ الذي بين مركز العالم ومركز الفلك المدير ونتوهّم قطر فلك التدوير بعد مفارقته للبعد الأبعد خارجاً على استقامة حتّى ينتهي إلى هذه النقطة فيسمّي* هذا الخطّ الخطّ المدير لقطر* فلك التدوير لأنّ فلك التدوير ينتقل أبداً على محيط فلك الحامل ويكون قطره المذكور* على استقامة الخطّ الذي يخرج من مركزه إلى

النقطة المذكورة وقد تسمّي* الدائرة التي يحدّها* طرف هذا الخطّ الفلك الخارج المركز أيضاً. [٢٩٦] فتكون* المراكز الثلاثة* في جهة واحدة من مركز العالم وهي التي تلي* البعد الأبعد أوّلها المركز* المدير لقطر فلك التدوير ويليه مركز* المدير للفلك الحامل ويليه مركز الفلك الحامل وأبعاد ما بينهما متساوية. [٢٩٧] وإذا تحرّك الفلك المدير لفلك الحامل مركز الفلك الحامل فإنّه يدور* في حركته حتّى (٢٨أ/ك) ينطبق على المركز المدير لقطر فلك التدوير ثمّ يفارقه* وينتهي في عودته إلى حيث كان أوّلاً.

[٢٩٨] فأما حركة الفلك الحامل والفلك المدير له فإنّهما متضادّتان (ب١١٢/ي) كما قدّمنا من القول إلاّ أنّ الفلك المدير يتحرّك إلى خلاف توالي البروج فيدور في كلّ سنة شمسية دورة واحدة فيعود قطره المارّ بمركز الفلك الحامل المدير له من مطابقة الخطّ الأوّل الذي عليه يكون البعد الأبعد والبعد الأقرب وعليه أيضاً تكون* مراكز الأفلاك إلى مطابقته أيضاً على الوضع الأوّل في كلّ سنة دفعة واحدة.

[٢٩٩] وأما الفلك الحامل المدير لفلك التدوير فإنّه يتحرّك إلى توالي البروج فيدور في هذه المدة من الزمان أعني السنة الشمسية دورتين حتّى أن الخطّ المدير لقطر الفلك التدوير يطابق الخطّ الذي يصل بين مركز* الفلك الحامل ومركز المدير له في هذه السنة دفعتين. [٣٠٠] وتلك المطابقة تكون* على الخطّ الذي عليه مراكز الأفلاك تارةً مما* يلي البعد الأبعد وتارةً مما يلي البعد الأقرب كمثل الحال في فلك القمر. [٣٠١] (ب٢٨/ك) فيعرض من هذه الحركة أن يكون مركز فلك التدوير ينتقل من مسامتة نقطة من الدائرة المائلة إلى أن يعود إليها في سنة واحدة شمسية ويكون* قد انتقل على محيط الفلك الحامل دفعتين.

[٣٠٢] ويقال إنّه يكون أيضاً* أبداً مسامتاً لموضع الشمس الوسط وذلك أنّه يوجد جرم الكوكب بالأرصاد المتّصلة يتباعد عن موضع الشمس الوسط متقدّماً عنه ومتأخّراً أبعاداً متساويةً في حركته على استقامة بالاضافة إلى فلك البروج وفي حركته راجعاً.

[٣٠٣] وهذه الحركة أعني الرجوع والاستقامة هي التي تخصّ فلك

التدوير. [٣٠٤] وحركة الكوكب في فلك التدوير* إنّما يتباعد* بها في الرجوع والاستقامة أعني من أعلى* فلك تدويره* وأدناه بعداً متساوياً* عن القطر وحده الذي يحدّ بطرفيه وسط الرجوع والاستقامة فقط فهذا القطر أبداً وهو المتّصل بالخطّ المدير مسامت لموضع الشمس فلذلك يكون فلك التدوير أبداً مسامتاً لموضع الشمس.

[٣٠٥] وجرم الكوكب يتحرّك حول مركز فلك* (١٢٩/ك) التدوير فيسامت* بحركته على ما سنبيّن أجزاء دائرة البروج أما إذا كان في أعلى فلك تدويره فإنّه يتحرّك على توالي البروج وأما إذا كان في أدناه فإنّه يتحرّك على خلاف توالي البروج بالضدّ من الحال التي للقمر إلاّ أنّ حركته مع ذلك تسبق* حركة فلك تدويره المستوية التي على توالي البروج. [٣٠٦] فيعرض من اختلاف الحركتين الحال التي تسمّي الرجوع وذلك أنّه إذا كان في أدنى فلك تدويره فحركته* في هذا الموضع على خلاف توالي البروج وجملة فلك التدوير بحركة* فلك الحامل على توالي البروج فحركته* تستبق حركة فلك التدوير أعنى أنه إذا تحرّك فتجاوز من دائرة البروج قوساً* مقدارها على طريق المثال* جزء واحد* متأخّراً ردّه فلك التدوير بالحركة العرضية التي للفلك الحامل بمقدار* نصف خزء فيُرى قد تأخّر قوساً من دائرة البروج مقدارها نصف خزء فيكون* باجتماع ذلك متأخّراً (ب٢٩/ك) فيسمّي ما دام بهذه الصفة راجعاً. [٣٠٧] وإذا كانت حركته في أعلى فلك التدوير وهي على توالي البروج كانت الحركتان جميعاً على توالي البروج واجتمع منها حركة واحدة فيسمّي حينئذ مستقيماً.

[٣٠٨] وأما الفلك المائل فإنّ ميله ميل يسير إلاّ أنّه ليس باق* على حاله بل يتغيّر بأن ينتقل جميع السطح فيقرب من سطح فلك البروج ولا يزال كذلك إلى أن ينطبق على سطح فلك البروج ثمّ يفارقه ويميل إلى الجهة الأخرى حتّى ينتهي إلى مثل ذلك المقدار من الميل ثمّ يعود كذلك أيضاً.

[٣٠٩] وأما قطر فلك التدوير الذي طرفاه البعد الأبعد والبعد الأقرب لفلك التدوير فليس يكون أبداً في سطح الفلك الخارج المركز ولا يلزم أبداً وضعاً واحداً بل يكون متحرّكاً على محيط دائرة صغيرة كأنّها عند طرفه* ويقود معه سطح فلك التدوير وهذه الدائرة قائمة على سطح

الفلك الخارج المركز على زوايا قائمة ومركزها فيه وتسمّي* حركة هذا القطر حول هذه الدائرة (٣٠١/ك) ميل فلك التدوير.

[٣١٠] والقطر أيضاً المقاطع* لهذا القطر (١١٢١/ي) على زوايا قائمة ليس هو أيضاً لازماً لوضع* واحد بل يتحرّك على دائرة صغيرة كأنّها عند طرفه وهي قائمة على سطح فلك البروج على زوايا قائمة وهذه الحركة تسمّي انحراف فلك التدوير وهي مساوية في السرعة للحركة الأولى وبحسب هاتين الحركتين وحركة فلك المائل يكون مجاز الكوكب* في العرض.

[٣١١] فإذا تحرّك فلك التدوير على محيط الفلك الخارج المركز فانتقل من البعد الأبعد ذاهباً إلى نقطة التقاطع ابتدأ الفلك المائل بالحركة نحو فلك البروج فإذا بلغ فلك التدوير إلى نقطة التقاطع انطبق الفلك المائل على سطح فلك البروج. [٣١٢] فإذا انتقل أيضاً فلك التدوير إلى ناحية البعد الأقرب انتقل أيضاً الفلك المائل ففارق سطح فلك* البروج ذاهباً بالبعد الأبعد إلى الجهة الأخرى وهي في هذه الحال جهة الشمال وانتقل البعد الأقرب إلى جهة الجنوب فإذا صار فلك* التدوير على البعد الأقرب من الفلك الخارج المركز انتهي ميل الفلك المائل.

[٣١٣] فإذا تحرّك أيضاً فلك التدوير إلى ناحية النقطة الأخرى من نقطتي التقاطع تحرّك أيضاً الفلك المائل عائداً إلى ناحية سطح فلك البروج فإذا انتهي إلى نقطة التقاطع انطبق أيضاً الفلك المائل على فلك البروج. [٣١٤] وإذا تحرّك من نقطة التقاطع إلى ناحية البعد الأبعد فارق أيضاً الفلك المائل سطح فلك* البروج عائداً إلى الجهة الأولى وهي جهة الجنوب فيكون مركز فلك التدوير إنّما يميل أبداً عن دائرة البروج إلى جهة الجنوب فقط.

[٣١٥] وأما ميل فلك التدوير عن سطح الفلك الخارج المركز المائل وهي حركة قطره حول الدائرة الصغيرة فإنّ ابتداءه يكون عند كون فلك التدوير في البعد الأبعد من الفلك* الخارج المركز فيفارق هذا القطر سطح الفلك الخارج المركز (ب٣٠/ك) وينتقل نحو الشمال ويقود معه سطح فلك التدوير ثمّ يعود أيضاً إلى سطح الفلك الخارج المركز ثمّ يفارقه نحو الجنوب

ثمّ يعود إليه أيضاً كذلك دائماً.

[٢١٦] فأما* القطر المقاطع لهذا القطر فإنّه يبتدئ بالمفارقة* عند كون فلك التدوير على نقطة التقاطع الرأس* وهي الجوزهر* فيتحرّك* نحو الشمال على محيط الدائرة الصغيرة القائمة على دائرة البروج على زوايا قائمة.

[٢١٧] وحركة هذين القطرين حول لدائرتين الصغيرتين ليست حول مركزها بل حول مركز خارج كمثل الحال في حركة الكوكب في الطول وعودات هذين القطرين حول لدائرتين بحسب الحركة في الطول. [٢١٨] وحركتها* في الأرباع أيضاً بحسبها حتّى إذا تحرّك الكوكب في الطول ربعاً من أرباع دائرته فعاد إلى نقطة الجوزهر أو إحدى النهايتين الشمالية والجنوبية تحرّك هذين القطرين ربعاً من الدائرة الصغيرة.

[٢١٩] فالحركات التي تخصّ كوكب عطارد تسع حركات*: حركة جميع فلكه على توالي البروج حول مركز فلك البروج وعلى قطبيه وهي في كلّ مئة سنة (٢١١/ك) جزء واحد وتسمّي* هذه الحركة حركة الأوج ؛ وحركة الفلك الخارج المركز المدير لفلك الحامل على خلاف توالي البروج وهي في كلّ سنة دورة واحدة وتسمّي* حركة المدير*؛ وحركة الفلك الحامل على توالي البروج وعلى قطبيه الثابتين الخاصّين به وتسمّي* هذه الحركة حركة مركز فلك التدوير وبحسب مسامتة هذه الحركة لدائرة البروج تكون* الحركة التي تسمّي حركة الطول* ؛ وحركة فلك التدوير حول مركزه وعلى قطبيه الثابتين المحرّك بها جرم الكوكب وتسمّي* حركة الاختلاف ؛ وحركة الخطّ المدير لقطر فلك التدوير حول النقطة الثابتة وتسمّي* حركة معدّل المسير؛ وحركة جميع سطح الفلك المائل إلى سطح فلك البروج وميله عنه وتسمّي* حركة الميل ؛ وحركة قطر فلك التدوير أعني المارّ بالبعدين البعد الأبعد والبعد الأقرب* على الدائرة الصغيرة مائلاً إلى الشمال وإلى الجنوب (ب١١٢/ي) وتسمّي* ميل فلك التدوير ؛ وحركة قطر فلك التدوير المقاطع لهذا القطر حول الدائرة الصغيرة إلى الشمال وإلى الجنوب أيضاً وتسمّي* انحراف فلك التدوير.

[٢٢٠] ونتوهّم خطّاً (ب٢١/ك) يخرج من مركز العالم إلى مركز الكوكب وينتهي على استقامة إلى سطح الفلك الأعلى ونتوهّم أيضاً دائرة

عظيمة تخرج من قطبي فلك البروج وتمرّ بطرف الخطّ فهذه الدائرة تقطع* محيط دائرة البروج فالنقطة التي تقطعها عليها هي التي تسمّي* موضع الكوكب من دائرة البروج والقوس من هذه الدائرة التي بين طرف الخطّ وبين نقطة التقاطع هي التي تسمّي* عرض الكوكب. [٣٢١] فهذه الجملة هي اقتصاص حال أفلاك عطارد وهذه صورة الدوائر التي ذكرناها.

القول على فلك الزهرة

[٣٢٢] فلك الزهرة جسم كريّ مستدير يحيط به سطحان متوازيان مركزهما مركز العالم يماس الأعلى منهما فلك الشمس ويحيط الأدنى منهما* بفلك عطارد وفي سطحه* الأعلى يكون الفلك المثّل بفلك البروج وهذا الفلك يتحرّك على قطبي فلك البروج حركة بطيئة على توالي البروج كمثل حركة فلك عطارد.

[٣٢٣] وينفصل من هذه الكرة كرة* أخرى مستديرة* يحيط بها سطحان متوازيان مركزهما وهو مركز الكرة خارج عن مركز العالم وسطحه الأعلى يماس السطح الأعلى من الفلك الأوّل على نقطة واحدة وهي تتحرّك* على قطبين ثابتين لها* حركة مستوية* إلى توالي البروج ويسمّي هذا الفلك الفلك الحامل.

[٣٢٤] وينفصل من هذه الكرة كرة مستديرة بين سطحيها* المتوازيين هي فلك التدوير وهي تتحرّك* حول مركزها على قطبين ثابتين لها وجرم الكوكب أعني الزهرة مركوز* في جسم هذا الفلك.

[٣٢٥] فإذا تحرّك الفلك الحامل حدثت* من حركته بنقطة مركز فلك التدوير دائرة متوهّمة ونتوهّم* فيها (٣٢١/ك) مثل ما توهّمنا في أفلاك عطارد فتحدث* في أفلاك الزهرة دوائر كمثل الدوائر التي حدثت في أفلاك عطارد أعني الدوائر الموازية للفلك الحامل والأفلاك المائلة.

[٣٢٦] ونتوهّم* خطّاً مستقيماً* يخرج من مركز العالم إلى مركز الفلك الحامل وينتهي إلى محيطه من الجهتين ويكون طرفاه البعد الأبعد* والبعد الأقرب. [٣٢٧] ونتوهّم* أيضاً قطر فلك التدوير على استقامة هذا الخطّ في البعد الأبعد فإذا تحرّك الفلك الحامل فانتقل قطر فلك التدوير فإنّه تسامت نقطة غير مركز الفلك الحامل ويكون بعدها من مركز الفلك الحامل كبعد هذا المركز من مركز العالم.

[٣٢٨] فإذا توهّمنا خطّاً مستقيماً يصل بين هذه النقطة ومركز فلك

التدوير كان على استقامة ذلك القطر وهذا الخطّ يسمّي الخطّ المدير لقطر فلك التدوير وتسمّي* النقطة المركز المدير وتسمّي* مركز معدّل المسير.

[٣٢٩] فالفلك الأوّل يتحرّك على قطبين مسامتين لقطبي فلك البروج فينقل جميع أفلاك الزهرة وينقل سطح الفلك المائل وبعده الأبعد وجميع ما فيه إلى توالي (ب٣٢/ك) البروج الحركة البطيئة وهي في كلّ مائة سنة جزء واحد.

[٣٣٠] والفلك الحامل يتحرّك على قطبيه الثابتين الخاصّين* به حول مركزه حركة مستديرة* إلى توالي البروج ويحرّك* بحركته فلك التدوير وينقل جرم الكوكب أيضاً* بانتقال فلك التدوير.

[٣٣١] ومركز الفلك الحامل ثابت غير منتقل ما سوى النقلة البطيئة الأولى بخلاف الحال في كوكب عطارد وفلك التدوير أيضاً يتحرّك على قطبيه حول مركزه فيحرّك جرم الكوكب أعني الزهرة حول مركزه.

[٣٣٢] ويحدث من حركة مركز جرم الكوكب دائرة مركزها مركز فلك التدوير تسمّي* أيضاً فلك التدوير وهذه الدائرة مائلة على سطح الفلك الخارج المركز كمثل الحال في فلك عطارد وحركة مركز الزهرة أبداً على محيطها أما إذا كان في أعلاها فحركته على توالي البروج وإذا كان في أدناه فحركته على خلاف توالي البروج.

[٣٣٣] فأما قطر فلك التدوير الذي طرفاه (١١٤١/ي) البعد الأبعد والبعد الأقرب فإنّه يتحرّك على دائرة صغيرة يقود* معه سطح فلك التدوير وكذلك القطر (٢٤١/ك) المقاطع له على زوايا قائمة كمثل* الصورة في فلك عطارد ما خلا أن المفارقة التي تكون* عن* سطح الفلك الخارج المركز تكون* عند البعد الأقرب من الفلك الخارج المركز وهي* هناك من البعد الأبعد والمفارقة* من القطر المقاطع تكون* هناك من إحدى* نقطتي التقاطع وهاهنا من النقطة المقابلة لها الذنب*.

[٣٣٤] وحركة هذين القطرين حول الدائرتين الصغيرتين ليست حول مركزهما* لكن* حول مركز خارج كمثل الحال في حركة الطول وعودات* القطرين بحسب عودات الطول.

[٣٣٥] فأما حركة الفلك المائل فمثل حركتها في كوكب عطارد إلاّ أنّ

ميلها إذا كان مركز* فلك التدوير على البعد الأبعد وعلى البعد الأقرب من الفلك الخارج المركز يكون إلى جهة الشمال فيكون ميل مركز فلك* التدوير أبداً إلى جهة الشمال بخلاف الحال في فلك عطارد.

[٣٢٦] فالحركات التي تخصّ كوكب الزهرة ثمان حركات*: حركة جميع فلكه* على توالي البروج الحركة البطيئة ؛ وحركة الفلك الحامل إلى توالي البروج ؛ والحركة* العارضة من هذه الحركة باضافتها إلى (ب٤٣/ك) دائرة البروج وهي حركة الطول ؛ وحركة فلك التدوير حول مركزه ؛ وحركة الخطّ المدير لقطر فلك التدوير ؛ وحركة قطر* فلك التدوير المائل؛ وحركة قطر فلك التدوير المقاطع له المنحرف* ؛ وحركة جميع سطح الفلك المائل.

[٣٢٧] فهذه الجملة هي حال أفلاك الزهرة وإنّما اختصرنا القول في حوال هذا* الكوكب لأنّ جميع حركاته وصورة أفلاكه شبيهة أكثرها بأفلاك عطارد فقد أتينا على وصفها في ذكر أفلاك عطارد فجميع* ما لم نورده* في هذه الجملة إنّما هو مقتصر به على أنّه مشروح* فيما تقدّم وهذه صورة أفلاك الزهرة (١٤١/ك).

القول على أفلاك الكواكب العلوية

[٣٣٨] أفلاك الكواكب العلوية الثلاثة أعني زحل والمشتري* والمريخ متشابهة في جميع أحوالها وعدد أفلاكها وأنواع حركاتها على ما شرحه بطلميوس في كتابه في التعليم* ولذلك لما* كانت لا تختلف* كثير اختلاف جعلنا القول في جميعها قولاً واحداً. [٣٣٩] وهو أنّ لكلّ واحد من هذه الكواكب كرة تخصّه* يحدّه سطحان كريان متوازيان مركزهما مركز العالم وتحيط* كلّ كرة بالتي تليها. [٣٤٠] فأوّلها فلك زحل سطحه* الأعلى يماس* كرة الكواكب الثابتة وسطحه الأدنى يماس محدّب* كرة المشتري وكذلك فلك المشتري يماس* بسطحه الأعلى* فلك زحل وسطحه الأدنى يماس فلك المريخ وكذلك فلك* المريخ سطحه الأعلى يماس فلك المشتري وسطحه الأدنى يماس فلك الشمس.

[٣٤١] وكلّ واحد من هذه الأفلاك يتحرّك على قطبين مسامتين لقطبي فلك البروج حركة بطيئة كمثل حركتي الزهرة وعطارد. [٣٤٢] وتنفصل* من كلّ واحد من هذه الأفلاك كرة خارجة المركز يحيط (ب٤١/ك) بها* سطحان كريان مركزهما مركز الكرة وهي تتحرّك* حول مركزها وعلى قطبين ثابتين لها حركة مستوية* على توالي البروج وتسمّي* هذه الكرة الفلك الحامل.

[٣٤٣] وينفصل من هذا* الفلك في كلّ واحد من الكواكب الثلاثة كرة مستديرة بين سطحيه المتوازيين كمثل ما قدّمنا فيما مضى تسمّي* كلّ واحدة* منها فلك التدوير لذلك الكوكب وهي تتحرّك* حول مركزها* وعلى أقطابها* حركة مستديرة. [٣٤٤] وجرم* كلّ واحد من الكواكب الثلاثة مركوز في جسم* فلك تدويره ويتحرّك بحركته.

[٣٤٥] فإذا تحرّك الفلك الحامل حركته التي تخصّه* وحرّك فلك التدوير رسمت نقطة مركزه دائرة متوهّمة تسمّي* أيضاً الفلك الحامل وإذا توهّمنا سطحها قاطعاً للعالم حدثت الدوائر المائلة والدوائر الموازية للفلك

الحامل كمثل الصورة في فلكي عطارد والزهرة إلاّ أنّ الدوائر المائلة (ب١١٤/ي) في هذه الأفلاك ثابتة على حال واحدة* ولا تنتقل* ولا تتغيّر*.

[٣٤٦] ونتوهّم* خطّاً مستقيماً يخرج من مركز العالم إلى مركز الفلك الحامل ويخرج على استقامته* (٤٢١/ك) إلى محيط الفلك المائل من الجهتين فيقطع محيط الفلك الحامل على نقطتين فتكون* إحديهما البعد الأبعد والأخرى البعد الأقرب. [٣٤٧] ونقطة البعد الأبعد من كلّ واحد* من هذه الخطوط أبداً في جهة الشمال ونقطة البعد الأقرب أبداً في جهة الجنوب* لأنّ ميل أفلاك هذه الكواكب المائلة لا يتغيّر ولا ينتقل وهي مائلة على وضع واحد بعدها الأبعد إلى جهة الشمال وبعدها الأقرب إلى جهة الجنوب.

[٣٤٨] فنفرض قطر فلك التدوير على استقامة هذا الخطّ فإذا تحرّك الفلك الحامل ونقل فلك التدوير انتقل قطره عن مسامتة ذلك الخطّ فحينئذ يوجد مسامتاً لنقطة أخرى غير مركز* الفلك الحامل ومركز العالم كمثل الحال في فلك الزهرة. [٣٤٩] وإذا وصل بين تلك النقطة وبين مركز فلك التدوير فيسمّي هذا الخطّ خطّ المدير لقطر فلك التدوير الكوكب كمثل الحال في فلكلي الزهرة وعطارد.

[٣٥٠] فالفلك الأوّل من أفلاك كلّ واحد من* الكواكب الثلاثة يتحرّك على قطبيه المسامتين لقطبي* فلك البروج الحركة (ب٤٢/ك) البطيئة وهي في كلّ مائة سنة جزء واحد ويتحرّك* بحركته جميع ما يحيط* به من أفلاك كوكبه وينقل* جميع سطح الفلك المائل والبعد الأبعد ومركز الفلك الحامل.

[٣٥١] والفلك الحامل يتحرّك حول مركزه الحركة المستوية التي على توالي البروج ويحرّك بحركته فلك التدوير وينتقل مع جملة فلك التدوير جرم* الكوكب.

[٣٥٢] وفلك التدوير يتحرّك حول مركزه فيحرّك جرم الكوكب فتحدث* من حركة مركز الكوكب دائرة متوهّمة مركزها مركز فلك التدوير تسمّي* أيضاً فلك التدوير وهذه الدائرة مائلة على سطح الفلك الخارج المركز كمثل الحال في الكوكبين المتقدّمين . [٣٥٣] ومركز كلّ واحد من الكواكب يتحرّك على محيط هذه الدائرة أما إذا كان في أعلاها فعلى* توالي

البروج وإذا كان في أدناها فإلى خلاف توالي البروج.

[٢٥٤] وقطر هذه الدائرة الذي طرفاه البعد الأبعد والبعد الأقرب لفلك التدوير فإنّه يتحرّك أيضاً على دائرة صغيرة فيميل تارةً إلى ناحية الشمال وتارةً (١٥٣/ك) إلى ناحية الجنوب ويقود معه سطح فلك التدوير.

[٢٥٥] ومفارقة القطر لسطح الفلك* الخارج المركز تكون* عند كون مركز فلك التدوير عند نقطة التقاطع الشمالية أعني الرأس* وإذا فارق مال أوّلاً إلى الشمال. [٢٥٦] وحركة هذه الأقطار حول الدوائر الصغار ليست حول مراكزها بل حول مراكز أخرى خارجة كمثل الحال* في حركة الطول. [٢٥٧] وعودات الأقطار حول الدوائر الصغار مساوية الأزمان للعودات في دائرة البروج وكذلك زمان حركاتها في الأرباع مطابقة* لحركة الطول في أرباع دائرة البروج.

[٢٥٨] فالحركات التي تخصّ كلّ واحد من الكواكب العلوية ستّ حركات: حركة* جميع فلكه الحركة البطيئة على توالي البروج ؛ وحركة الفلك الحامل على قطبيه المدير لفلك التدوير على توالي البروج ؛ والحركة التي تعرض من هذه الحركة بالاضافة إلى دائرة البروج وهي حركة الطول* ؛ وحركة فلك التدوير حول مركزه ؛ وحركة الخطّ المدير لقطر فلك التدوير (ب٢٥/ك) حول المركز المدير ؛ وحركة قطر فلك التدوير المائل.

[٢٥٩] فهذا* هو اقتصاص حال أفلاك الكواكب العلوية وهي جملة تفصيلها كمثل التفصيل في الكوكبين الأوّلين ولذلك اقتنعنا* بالايجاز* عن التكثير. وهذه صورة أفلاكها وهي لكلّ واحد من الكواكب* الثلاثة.

القول على فلك الكواكب الثابتة

[٣٦٠] (ب٢٦/ك) فلك الكواكب الثابتة كرة واحدة مستديرة يحيط به سطحان كريان مركزهما وهو مركز (١١٥ا/ي) الكرة مركز العالم والسطح الأعلى منهما يماس الفلك الأعظم المحيط بجميع الأفلاك المتحرّك الحركة السريعة والسطح* الأدنى منهما مماس لفلك زحل. [٣٦١] وهذا الفلك يتحرّك من جهة المغرب إلى جهة المشرق على* توالي البروج على قطبين ثابتين حركة بطيئة وهي في كلّ مئة سنة جزء واحد من الأجزاء* التي بها الدائرة ثلاثمئة وستّون جزءاً وقطباهما هما* قطبا دائرة البروج التي رسمتها الشمس على ما ذكر بطلميوس أنّه وجده* بأرصاده وأرصاد من تقدّمه.

[٣٦٢] وجميع الكواكب الثابتة مركوزة في جسم الفلك وغير مفارقة لمواضعها ولذلك لا تتغيّر* أوضاعها بعضها عند بعض لكنّها تنتقل* بجملتها فتتحرّك* بحركة فلكها الحركة البطيئة التي له إلى توالي البروج. [٣٦٣] وكلّ واحد منها* (٢٧ا/ك) يحدث مركزه بحركة هذا الفلك* دائرة وهذه* الدوائر كلّها متوازية وقطباها قطبا دائرة البروج وكلّ واحد منها يتحرّك أبداً على محيط دائرته وغير مفارقة لها.

[٣٦٤] وكلّ واحد من هذه الكواكب فحركته بإضافة إلى دائرة البروج وموضعه منها كمثل الحال في سائر الكواكب وهو أن نتوهّم دائرة خارجة من قطبي فلك البروج وتمرّ بمركز الكوكب فهي تقطع العالم وتقطع* دائرة البروج التي في سطح الفلك الأعلى وتقطع* أيضاً دائرة البروج التي في سطح فلك البروج على نقطة. [٣٦٥] فتلك النقطة هي موضع الكوكب من دائرة البروج التي في سطح فلك البروج والنقطة المسامتة لها من دائرة البروج التي في سطح الفلك الأعلى هي* موضعه المحقّق في الطول وحركة تلك النقطة المتحرّكة* بحركة الدائرة المارّة بالقطبين هي* حركة الكوكب في الطول.

[٣٦٦] والخطّ الذي يخرج من مركز العالم إلى مركز الكوكب وينتهي إلى سطح الفلك الأعلى هو أيضاً في سطح تلك الدائرة ونقطة نهايته على

محيط (ب٣٧/ك) تلك الدائرة وتلك النقطة هي موضع الكوكب من سطح الفلك الأعلى. [٣٦٧] والقوس من الدائرة التي تمرّ بالقطبين التي بين هذه النقطة وبين موضع الكوكب من دائرة البروج هي* عرض الكوكب.

[٣٦٨] وجميع فلك الكواكب* الثابتة يُقسم بستّ* دوائر تخرج* من قطبيه فتقسمه* بإثنى عشر قسماً متساويةً هي الأقسام التي ذكرناها فيما* تقدّم من القول على جملة الفلك كلّ قسم منها بين نصفي دائرتين من هذه الدوائر الستّ*. [٣٦٩] وتقسم* هذه الدوائر دائرة البروج أيضاً التي في سطح هذا الفلك بإثنى عشر قسماً متساويةً فالأقسام الإثنى* عشر من جميع الفلك تسمّي* بروجاً. والقسي التي فيها من دائرة البروج تسمّي* أيضاً بروجاً.

[٣٧٠] وكلّ واحد من أقسام جميع الفلك يسمّي كما قلنا فيما تقدّم باسم الشكل من أشكال الكواكب التي فيه* وتسمّي* القسي من دائرة البروج بأسماء* تلك البروج والقسي من دائرة البروج التي في سطح الفلك الأعلى أيضاً بأسماء* تلك البروج وأقسام جميع الفلك الأعلى تسمّي بأسماء* تلك الأقسام فأقسام* جميع أفلاك الكواكب أيضاً تسمّي* بأسماء* تلك الأقسام.

[٣٧١] (١٢٨/ك) وفي هذا الفلك من الكواكب الثابتة عدد كثير مختلف الأقدار مثبت في جميع جسم هذا الفلك إلاّ أنّ منها ما قد رُصدت* مواضعه وعُلمت أمكنته من* دائرة البروج وعُرفت عروضه وذلك عدد مبلغه على ما ذكر بطلميوس في كتابه في التعليم ألف وإثنان وعشرون كوكباً قسمها بأقسام جمع* في كلّ قسم منها ما كان منها متساوياً في العظم. [٣٧٢] فذكر أن الذي يجتمع في العظم الأوّل منها وهو أعظمها خمسة عشر كوكباً وفي العظم الثاني خمسة وأربعون كوكباً وفي العظم الثالث مائتان وثمان كواكب وفي العظم الرابع أربعمائة وأربعة وسبعون كوكباً وفي العظم الخامس مائتان وسبعة عشر كوكباً وفي العظم السادس تسعة* وأربعون كوكباً والخفية* تسع كواكب والسحابية خمسة* كواكب وذكر مواضع هذه من دائرة البروج وعروضها (١١٥ا/ي) بالاضافة إلى دائرة البروج. [٣٧٣] وحصرها أيضاً* بطريق ينضبط به عددها ويُعرف به مواضعها* من سطح الفلك وذلك أنّه نظر

إلى ما يتشكّل من الأشكال باجتماع كلّ جماعة من هذه (ب٣٨/ك) الكواكب فسمّي تلك الجملة باسم تلك الصورة فجمع* هذه الكواكب المعدودة في ثمانية واربعين* صورة يسمّي* كلّ واحد منها باسم ما يشبهه من الحيوان وغيره فصار جميع المرصود من الكواكب الثابتة وهو ما يشتمل عليه الثمانية والأربعون صورة والعدد المذكور.

[٣٧٤] فأما ما سوى ذلك فهو* الذي* في غاية الصغر وما لعلّة قربه* من القطب الجنوبي خفي عن أهل* المساكن الشمالية فإنّه غير معدود ولا معروف المواضع من دائرة البروج وهذه صورة فلك* الكواكب الثابتة*.

القول على الفلك الأعلى

[٢٧٥] (ب٣٩/ك) أما* الفلك الأعلى المحيط بجميع الموجودات فإنّه محيط بكرة الكواكب الثابتة مماس لها وهو يتحرّك على قطبيه اللذين هما قطبا العالم الحركة السريعة من المشرق إلى المغرب ويتحرك بحركته جميع أفلاك الكواكب على ما بيّناه في أوّل هذه المقالة وليس فيه شيء من الكواكب.

[٢٧٦] وجميع الدوائر المائلة* التي لسائر الكواكب والدوائر التي تمرّ بقطبي دائرة البروج التي توجد منها* عروض الكواكب وجميع الدوائر العظام التي تقدّم في أوّل المقالة ذكرها ودوائر البروج التي إليها* تقاس* حركة جميع الكواكب وأطراف الخطوط التي تخرج* من مركز العالم وتمرّ بمراكز* الكواكب فإنّها متوهّمة كلّها في سطح الفلك الأعلى المحيط بجميع الأفلاك وإليه يقاس جميع أحوال الكواكب في الحركات والاتّصلات والاجتماعات والاقترانات. [٢٧٧] أما سائر الحركات وأوضاع الاتّصلات فبالقياس إلى دائرة البروج التي في سطح الفلك الأعلى بأوضاع النقط منها التي (ا٣٩/ك) بيّنا أنها مواضع الكواكب في الطول.

[٢٧٨] وأما أبعاد ما بين الكواكب بالحقيقة* فمن الدوائر العظام التي تمرّ بمركز العالم وبطرفي الخطّين من سطح الفلك الأعلى* اللذين هما موضعا الكوكبين.

[٢٧٩] وأما* الاقترانات فبالدائرة التي تمرّ بقطبي فلك البروج وبطرف الخطّ الذي هو موضع الكوكب من سطح الفلك الأعلى إذا كانت مع هذه الحال تمرّ بطرف الخطّ الذي هو موضع الكوكب الآخر فإنّه يعرض من هذه الحال أن يكون موضعهما* من دائرة البروج نقطة واحدة لأنّ الدائرة التي تمرّ بموضعهما* دائرة واحدة وكذلك الاجتماعات التي للشمس والقمر بيّنا ذلك في موضعه من ذكر أفلاك القمر.

[٢٨٠] فجميع الحركات الموجودات في جميع أجزاء العالم على ما فهمناه بقدر اجتهادنا من أراء* أصحاب التعليم وأرصادهم المدونة وعلى* ما يوجد

بالاستقراء* بحسب ما ينتهي* إليه فحص من عني بهذا العلم وخاصةً بطليميوس فأنا على بحثه وأرصاده أعمدنا في جميع الحركات السماوية (٤٠١/ك) على* ما قدّمنا* من تفصيلها سبعة وأربعون حركة: [٣٨١] حركة الفلك الأعظم المحيط من المشرق إلى المغرب وهي الحركة السريعة ؛ وحركة فلك الكواكب الثابتة من المغرب إلى المشرق وهي الحركة البطيئة ؛ وثماني* عشر حركةً للكواكب* الثلاثة العلوية لكلّ واحد منها ستّ حركات ؛ وحركاتان للشمس*؛ وثماني* حركات للزهرة ؛ وتسع حركات لعطارد ؛ وستّ حركات للقمر ؛ وحركتان لما دون فلك القمر* وهما حركتا الثقل والخفة ؛ ومن جملة حركات كلّ واحد من الكواكب السبعة المتحيّرة حركة واحدة بالاضافة لا ذاتية وهي حركته التي توجد بالقياس إلى دائرة البروج.

[٣٨٢] ولما كان قولنا في كلّ واحد من الأفلاك متخصّصاً وتخطيطً* صورته مفردة* إنّما يُفهم منه ويُتصوّر فلك واحد واحد من الكواكب قائماً بنفسه ولا يكاد أن يُتصوّر من مجموعها جملة الأفلاك مجتمعة ودوائرها مقاطعة (٤٠١/ك) وأفلاكها متماسة وجب أن نخطّ صورةً نرسم فيها جميع أفلاك الكواكب مجتمعة على ترتيبها وبحسب هيئتها (١١٦١/ي) وعلى* ما شرحناه متفارقاً* من أمرها ليُعقل من جملتها* صورة العالم بجميع ما يحيط به وما يشتمل عليه. [٣٨٣] ولكي لا يكثر الأفلاك فيستعجم* على الناظر في الصورة نقتصر في كلّ فلك منها وضع واحد من أوضاعه ونكتفي به دالاً في حركته على سائر أوضاعه. [٣٨٤] ونفرض لتسهيل العمل جميع الدوائر المائلة لجميع* الكواكب في سطح واحد وأبعد* أبعادها أيضاً على خطّ واحد ومراكز أفلاكها الخارجة المراكز كلّها على ذلك الخطّ فيكون ذلك أبين في التمثيل ولا يؤثر ضرراً في أصول الصور. [٣٨٥] ونجعل أيضاً المركز الخارج المدير لأقطار أفلاك التداوير* للكواكب العلوية وكوكب* الزهرة واحداً ونجعله قريباً من مركز العالم ليكون أبين* إذ ليس الغرض في الصورة إلاّ (٤٢١/ك) المثال فقط وكذلك نجعل مركز الفلك الحامل لجميع الكواكب خلا عطارد واحداً وهذه* صورة الكلّ.

التعليق وخاتمتي المخطوطين

خاتمة مخطوط «ي»

تم الكتاب ولخالق الكلّ وواهب العقل الحمد والشكر* والمنة الصلوة على خير خلقه محمّد المصطفى وعلى آله وعشرته الطاهرين أجمعين وسلم.

تعليق وجدناه بخطّ الشيخ أطال الله بقاؤه في آخر هذه المقالة فنقلناه كما وجدناه:

لّما كانت الكواكب أجساماً طبيعيةً وجب أن يكون كلّ واحد منها إنّما يصحّ أن يتحرّك بذاته حركة واحدة طبيعية ولّما كانت الكواكب من جوهر بسيط واجب أن يكون حركاتها مستوية متشابهة الحال كلّ واحد منها لا يختلف حركته ولّما كان المتحرّك بذاته المنتقل من مكان إلى غيره يفرع مكاناً ويملأ غيره لم يخل* من أحد أمرين أما أن يخلو* مكانه الذي كان فيه أو يملأه جسم آخر من الأجسام المحيطة به وليس يمكن أن يخلو* مكانه فإن تبدّلت الأجسام على الأمكنة التي تمرّ بها لزم من ذلك أن ينخرق الجسم أو الأجسام التي تتحرّك* فيها ولّما كان جسم السمآء من طبيعة لا يقبل الانفعال بطل أن يكون شيء من الكواكب يتحرّك على هذا الوجه فليس يبقى لحركتها وجه سوى أن يكون لكلّ واحد منها جسم مستدير يتحرّك على نفسه حركة مستديرة يكون جرم الكوكب ثابتاً في موضع من ذلك الجسم فيتحرّك الكوكب بحركته ولا يحتاج هو ولا الجسم المحرّك له إلى مكان أكثر من المكان الذي هما فيه ولّما كان كلّ واحد منها* يوجد* له حركات مختلفة وليس يصحّ أن يكون لذاته أكثر من حركة واحدة مستوية لزم من ذلك أن يكون لكلّ واحد منها محرّكات هي منقادة* لها وتابعة لحركاتها كلّ واحد منها يحرّكها نوعاً من تلك الحركات فمن أجل ذلك لزم أن يكون لكلّ واحد منها أو لبعضها أجسام أكثر من واحد تتحرّك* على أنفسها ولا تحتاج*

إلى مكان أكثر من مكانها وتتحرّك* بحركتها الكواكب المتّصلة بها فمن أجل جميع ذلك وجب أن يكون للكواكب أكر كثيرة محيطة بعضها ببعض ولكلّ واحدة منها حركة تخصّها على قطبين ثابتين لها ومحور ثابت تتحرّك* عليه ولا تحتاج* إلى أكثر من مكانها.

فصل آخر : المقدّمات التي عليها ينبني تركيب أفلاك* الكواكب وجميع الأجسام المتحرّكة حول العالم أربع : إحداها إنّ الجسم الطبيعي لا يتحرّك بذاته حركة* طبيعية أكثر من حركة واحدة ؛ والثانية إنّ الجسم الطبيعي لا يتحرّك حركة مختلفة أعني إنّه أبداً يقطع في الاستدارة في الأزمنة المتساوية مسافات متساوية ؛ والثالثة إنّ جسم السمآء جسم لا يقبل الانفعال ؛ والرابعة إنّ الخلاء ليس بموجود.

خاتمة مخطوط «ك»

تم القول والحمد لله ربّ العالمين وصلى الله على* سيّدنا محمّد وآله أجمعين ورضي الله عن الصحابة والتابعين والمسلمين أجمعين وكُتب هذا الكتاب من النسخة التي نُسخت* من نسخة الشيخ ابن القسم السميساطي بخطّه ذكر أنه نقلها من نسخة بخطّ مصنّف الكتاب الشيخ ابن علي الحسن بن الحسن ابن الهيثم وقابل عليها من أوّلها إلى آخرها في رجب من سنة ستّ وسبعين واربعمائة.

والنسخة المكتوبة منها* هذه النسخة عورض بها نسخة* الاصل المذكورة* وهي بخطّ الشيخ ابن علي الهيثم وصُحّح والحمد لله ربّ العالم وكُتب في رجب من السنة المذكورة.

طالع في هذا الكتاب المبارك عبد الله بن محمّد بن (؟).

جهاز التحقيق

لعنوان العزة لله: استعنت بالله (ك)
قول للحسن: قول الشيخ الفاضل ابي علي الحسن (ك)
[١] قال: ــ (ك)
يشرحون: تشرحون (ي، ك)
[٢] هي المتحركة: غير واضح في ك.
بتلك: وبتلك (ك)
[٣] تتم: يتم (ي)
لاهل: اهل (ك)
لا: + مستنكراً: (ي)
فهم: افهم (ك)
[٤] تقتضيه: يقتضيه (ي، ك)
[٥] اهملوها: اهملواها (ك)
لكن: ولكن (ك)
[٦] اذ: اذا (ك)
بياناً: ايضاً (ي)
[٧] فهمناه: فهمنا (ي)
[٩] واحدة: واحد (ي، ك)
تظهر: يظهر (ي، ك)
جسم كري: جسمي جزي (ي)
[١٠] جميع: ــ (ك)
من: + من (!)(ك)
تكون: يكون (ي)
[١١] التنقل: النقل (ي)
[١٢] الحركات: الاوضاع (ك)
نورده: يورده (ك)
صورة كلّيّة الشكل: صور الكلّيّة للشكل (ك)
بها: به (ي)
جزؤياتها: جزاياتها (ي)
تكون: يكون (ي)
[١٣] الكرة: الدائرة والكرة (ي)
[١٤] كرة: ــ (ك)
النقطة: النقط (ك)
خال: خالي (ي)
[١٥] تملأ: تمال (ي)

جهاز التحقيق

[١٦] المركز: بقدر ما له من القوة (ك)
فهو: هو (ك)
[١٧] نواحي: النواحي (ي)
[١٩] ان رفع:ارفع (ك)
مكانه: مكانها (ي)
حيث: _ (ك)
[٢٠] ذلك: _ (ي)
[٢١] تضاريس: تضار ليس (ي)
تعرض: يعرض (ك، ي)
تأثيرات حركات الاجرام: تأثيرات الاجرام (ي) ؛ تأثيرات حركات اجرام (ك)
بمخرج : بمخرجها (ي)
[٢٢] بضرب : لضرب (ك)
[٢٣] الاّ انّه : الّس ان المآء (ي)
وكانت : وكان (ي)
المركز : مركز (ي)
بينه : بينها (ي)
من المركز : من الارض الى المركز (ك)
[٢٤] متشعّثة : مشعّثة (ك)
وكان : فكان (ي)
من اجل : من اجلد (ك)
تكون : يكون (ي)
يحيط : محيط (ي)
[٢٥] تكون : يكون (ي)
التي تكون خارجةً عن المآء : _ (ك)
[٢٦] من : _ (ي)
صحيحة : صحيح (ك)
متساوية : متشابهة (ك)
منفرداً : مفرداً (ك)
[٢٧] بثقيل ولا خفيف : بخفيف ولا ثقيل (ك)
[٢٨] الطف منه : منه الطف (ي)
منه : _ (ي)
[٢٩] ينقسم : تنقسم (ي)
بقسمين : قسمين (ي)
[٣٠] كتشعّث : كتشعّت (ي)

جهاز التحقيق

[٢١] فهذا : وهذا (ك)
عوارض : العوارض (ك)
اجزائه : اجزائها (ي)
[٢٢] فيه : فيها (ي)
تعرض : يعرض (ي)
المؤتلف : المتلفة (ك)
في : هي (ي)
[٢٣] شعاعات : شبعاعات (ي)
ساخناً : سخناً (ي ، ك)
بها : لها (ك)
[٢٤] محيطاً : محيط (ي)
يحدّه : + ويحقّقه (ي)
الجسم : والجسم (ك)
فهذا : وهذا (ي)
[٢٦] بكليّته : بكليّة (ي)
يحيط : تحيط (ك)
متشابه : متشابهة (ي)
[٢٧] تسمّي : يسمّي (ي)
ويحرّك : وتحرّك (ك)
[٢٨] من ذلك : ــ (ي)
[٤٠] الى : ــ (ي)
احدها : احدهما (ي)
[٤٢] المحيط : ــ (ي)
الافلاك : المحيط (ي)
[٤٣] تنقسم : ينقسم (ي)
[٤٥] تسمّي : يسمّي (ي)
[٤٦] تسمّي : يسمّي (ي)
[٤٧] ايضاً : ــ (ك)
ولكلّ : لانّ كلّ (ي)
منها : ــ (ي)
[٤٨] واحدة : واحد (ي)
تنقسم : ينقسم (ي ، ك)
تخصّها : تخصّه (ي)
واحدة منها : واحد (ي)

جهاز التحقيق

[٤٨] موضوعها : موضوعه (ي ، ك)
ان شاء الله : ـ (ي)
[٤٩] تتحرّك : يتحرّك (ي)
ترسم : + في (ي)
متماستا : متماسةً (ي ، ك)
واحدة : واحد (ي)
[٥٠] كوكب : كوب (ي)
يحدث : تحدث (ك)
وبمركز : ومركز (ك)
ويقطع : وتقطع (ك)
الاعلى : الاعظم (ي)
واحدة : واحد (ي)
خاصّة : وخاصّة (ي)
[٥١] وكلّما : وكلّ ما (ي)
[٥٢] تسمّي : يسمّي (ي)
في جميع العالم : في الارض المعمورة وفي نسخة آخر في جميع العالم ما سوى النقطتين المسامتتين لقطبا العالم (ك)
هما : سقط من مخطوط «ك» النص التالي من «قطبا العالم» الى المكان المرسوم في فقرة ٦٢.
[٥٣] الدائرة : الدوائر (ي)
[٥٤] اجزاء : جزا (ي)
[٥٥] عشر : عشرة (ي)
يسمّي : تسمّي (ي)
فتكون : فيكون (ي)
[٥٧] وبدورة : بدورة (ي)
[٥٨] ونتوهّم : نتوهّم (ي)
فتقطع : فيقطع (ي)
وتقطع : ويقطع (ي)
[٥٩] وتقطع : ويقطع (ي)
[٦٠] تحدث : يحدث (ي)
تسمّي : يسمّي (ي)
الاستواء : الاستوي (ي)
الذي : ـ (ي)
[٦٢] وتسمّي : ويسمّي (ي)
[٦٢] تقسم : يقسم (ي ، ك)

جهاز التحقيق

[٦٢] وتحدّ : فتحدّ (ي)
تسمّي : يسمّي (ي)
[٦٤] الاعلى : الاعظم (ي)
رؤوس : رؤس (ي) ؛ الرؤس (ك)
والدائرة العظيمة...الارض : ـ (ي)
[٦٦] ولكلّ : فلكلّ (ي)
من : على (ك)
[٦٧] فيكون : فتكون (ك)
وتكون : ويكون (ي)
معتدلاً : معدّلاً (ك)
[٦٨] وتكون : ويكون (ي)
عليها : عنها (ي)
مائلاً : مائلة (ي ، ك)
عليها الى : عنها مما (ي)
[٦٩] ايّ : ـ (ك)
الافق : فانّما سمّيت دائرة نصف النهار (ي)
الذي : التي (ي)
[٧٠] تقسم : يقسم (ك)
فتحدث : فيحدث (ي)
والاخرى : صحّ والنقطة المقابلة (ك)
وتسمّي : ويسمّي (ي)
[٧١] تقسم : يقسم (ك)
فتحدث : فيحدث (ي)
تسمّي : يسمّي (ي)
[٧٢] الشمال...والمغرب : الشمالي والجنوبي والمشرقية والمغربية (ي)
والخطّ المستقيم...والمغرب : ـ (ي)
[٧٤] ونتوهّم : ويتوهّم (ي ، ك)
تكون : وهي يكون (ي)
وتقطع : ويقطع (ي)
[٧٥] تسمّي : يسمّي (ك)
وتانك النقطتان : وتلك النقطتين (ي)
ويسمّي : وتسمّي (ي ، ك)
[٧٦] او : وبين (ي)
[٧٨] وتسامتها : ويسامتها (ي ، ك)

جهاز التحقيق

[٨١] وطرفا : وطرف (ي)
[٨٢] وتتوهّم : ويتوهّم (ي ، ك)
تقطع : يقطع (ي ، ك)
بقسمين : بنصفين (ي)
[٨٣] دائرة...بنصفين : ـ (ي)
وتقسم : ويقسم (ك) ؛ ـ (ي)
[٨٣] تكون : يكون (ي)
الافق : الارض (ي)
[٨٤] القسي : القسمي (ي)
قسي : قسمي (ي)
تسمّي : يسمّي (ي ، ك)
[٨٥] كانت : كان (ي ، ك)
[٨٦] شبهاً : شبه (ك)
المقسومة : المقسوم (ي ، ك)
تكون : يكون (ي ، ك)
عدداً: ـ (ك)
كانت : كان (ي ، ك)
الدوائر : الدائرة (ك)
فوق : فوقما (ي)
وكانت : وكان (ي ، ك)
[٨٧] تتحرّك : يتحرّك (ي ، ك)
[٨٨] تلي : يلي (ي)
القوس : القسي (ك)
[٨٩] تقسم : يقسم (ي ، ك)
فتكون : فيكون (ي ، ك)
للقوس : للنفوس (!) (ي)
الافق : الارض (ي)
[٩٠] تقمع : تقطع (ي)
تقطعها : يقطعها (ي)
احديهما : احداهما (ك)
[٩١] تطلع : يطلع (ي)
النقط : النقطة (ي)
وتغرب : ويغرب (ي ، ك)
الغربية : ـ (ي)

[٩٢] تسمّي : يسمّي (ي)
لتلك : لتك (ي)
تطلع : يطلع (ي)
تسمّي : يسمّي (ي)
[٩٥] ما : ـ (ي)
جميعها ظاهراً : دائمة الظهور (ي)
وتكون : ويكون (ك) ؛ ـ (ي)
الخفي : الخفا (ي)
جميعها خفياً : دائمة الخفاء (ي)
وتكون : ويكون (ك) ؛ ـ (ي)
[٩٧] وتقسم : ويقسم (ي)
تسمّي : يسمّي (ي)
[٩٨] الاقليم : الاقاليم (ي)
فخطّ : بخطّ (ك)
والليل...ابداً : ويكون الليل والنهار ابداً في هذه الموضع (ي)
الاقليم : الاقاليم (ي)
[٩٩] مدار : + رأس (ي)
جميعه : جميعها (ي)
تنتهي : ينتهي (ي)
مدار : + رأس (ي)
[١٠٠] الدائرة : الدوائر (ي)
[١٠١] تقسم : يقسم (ي ، ك)
فمنهم...العمارة : فيجعل مبداها (ك)
ولا...من العمارة : ـ (ي)
[١٠٢] ويجعل : ونجعل (ي)
في الجنوب : ـ (ي)
الاستواء : + في جهة الجنوب (ي)
عشر : عشرين (ك)
وسدس : سدس (ي)
النهار : نهار (ك)
الجنوبي : الجنوبية (ي)
[١٠٥] الليل والنهار : النهار والليل (ك)
[١٠٦] وتكون : ويكون (ي ، ك)
تكون : يكون (ي)

[١٠٦] الافق : الارض (ي)
تكون : يكون (ي)
تتحرّك : يتحرّك (ي)
ترتفع : يرتفع (ي ، ك)
تكون : يكون (ي ، ك)
[١٠٧] حدثت : حدث (ي ، ك)
[١٠٨] ونتوهّم : ويتوهّم (ي ، ك)
فتحدث : فيحدث (ي ، ك)
تسمّي : يسمّي (ي)
يسمّيان : تسمّي (ي ، ك)
[١٠٩] تقطع : يقطع (ي)
[١١٠] ونتوهّم : ويتوهّم (ي)
فتقطع : فيقطع (ي)
فتسمّي : فيسمّي (ي)
[١١١] ونتوهّم : ويتوهّم (ي)
تقسمان : يقسمان (ي)
[١١٢] الربع : البيع (ي)
[١١٣] الربع : البيع (ي)
الخريف : الخريفي (ي ، ك)
[١١٤] الربع : البيع (ي)
تسامته : مسامتة (ك)
السنة الشمسية : ستة شمسية (ي)
[١١٥] الخاصّة : الخاصّ (ي)
تعود : يعود (ي)
[١١٦] تنتهي : ينتهي (ي)
[١١٧] تزال : يزال (ي)
تنتهي : ينتهي (ي)
[١١٨] تتحرّك : يتحرّك (ي)
تنتهي : ينتهي (ي)
تكون : يكون (ي)
تبريدها : تبزيدها (ي)
[١١٩] تعود : يعود (ي)
فيعرض : فيغرض (ي)
[١٢٠] ونتوهّم : ويتوهّم (ي)

جهاز التحقيق

[١٢٠] دائرتين عظيمتين : دائرتان عظيمان (ي)
تخرجان : يخرجان (ي)
فتقطعان : فيقطعان (ي)
بثلاث : ثلثة (ي)
فتقطعان : فيقطعان (ي ، ك)
ونتوهّم : ويتوهّم (ي)
دائرتين اخريين : دائرتان اخرتان (ي)
وتقطعان : ويقطعان (ي)
الصيفي : + بثلاث اقسام (ي)

[١٢١] ستّ : ستة (ي)
تقسم : يقتسم (ي) ؛ يقسم (ك)
تقسم : تقتسم (ي) ؛ يقسم (ك)
الستّ : الستّة (ي)
وتسمّي : ويسمّي (ي)
[١٢٢] ونتوهّم : ويتوهّم (ي)
الستّ : الستّة (ي)
قاطعةً : القاطعة (ي)
افلاك : الافلاك (ي)
[١٢٤] وتقسم : ويقسم (ك)
وتحدث : ويحدث (ي)
[١٢٥] وتنقسم : وينقسم (ي)
الستّ : الستّة (ي)
تسمّي : يسمّي (ي)
بروجاً : برجاً (ي)
[١٢٦] ويسمّي : وتسمّي (ي)
البرج : البروج (ي)
وتسمّي : ويسمّي (ك)
[١٢٧] اقسام : الاقسام (ي)
هذه : هذا (ي)
الربيعي : ــ (ك)
اذ : اذا (ك) ؛ اذن (ي)
[١٢٨] القاطعة : + النقطة (ي)
فيسمّي : فتسمّي (ك)
وتسمّي : ويسمّي (ي)
[١٣٠] البروج : البرج
[١٣١] تكون : يكون (ي)
البرج : البروج (ك)
[١٣٢] نزولها : تزاولها (ي)
ويخرج : ونخرج (ي)
[١٣٣] مارة : مادة (ي)
تقطع : يقطع (ي)
في الطول : ــ (ي)
[١٣٤] واحدة : واحد (ي)

جهاز التحقيق

[١٢٤] الستّ : الستّة (ي)
تقطع : يقطع (ي)
تلك: ذلك (ي)
[١٢٦] تقطع : يقطع (ي)
وتسمّي : ويسمّي (ي)
مدار : + رأس (ك)
[١٢٧] وبرأس : ورأس (ي)
وبرأس : واراس (ي)
وبرأس : ـ (ي)
[١٢٨] تقطع : يقطع (ي ، ك)
وتحصل : ويحصل (ي)
تسمّي : يسمّي (ي)
تلك : ـ (ك)
تحرّك : تحرّكت (ي)
تطلع : يطلع (ي)
منها : منه (ي)
[١٣٩] على دائرة : ـ (ي)
تقطع : يقطع (ي
تطلع : يطلع (ي ، ك)
تسمّي : يسمّي (ي)
[١٤٢] تخرج : يخرج (ي)
تقطعها : يقطعها (ي)
وتنتهي : وينتهي (ي ، ك)
فتحصل : فيحصل (ي)
[١٤٣] وكلّما : وكلّ ما (ي)
[١٤٤] فاعظم : ـ (ي)
الانقلاب : + الصيفي (ك)
قريبة : قريب (ي)
[١٤٥] وبرأس : وراس (ي)
تسمّي : يسمّي (ي)
والمقابلة : والمقابل (ي)
تسمّي : يسمّي (ي)
[١٤٦] وبرأس القوس فانّ القوس منها : شطب في « ك »
تسمّي : يسمّي (ي)

جهاز التحقيق

[١٤٦] الثور والعقرب : الحمل والثور (ي)
[١٤٧] السرطان والجدي : الاسد والسنبلة (ي ، ك)؛وفي « ك » شطب « السنبلة » وكتب في الهامش « الدلو »
[١٤٨] وبرأس : وراس (ي)
الاسد والدلو : السنبلة (ي) ؛ السنبلة والحوت (ك)
[١٥٠] دائرة : دوائر (ي)
[١٥١] تكون : يكون (ي)
[١٥٢] تكون : يكون (ي)
[١٥٢] جزءاً : درجة (ك)
الانقلاب : + الصيفي (ك)
نهار هذا : النهار لذلك (ك ؛ وفي الهامش نجد كلمة « هذا »)
[١٥٤] تقطع : يقطع (ي)
تقطعها : يقطعها (ي)
متساو : متساويتين (ي) ؛ متساويا (ك)
لانّه : + هو (ك)
تحرّكت : بحركت (ي)
[١٥٥] تنتهي : ينتهي (ي)
الثانية : الثابتة (ي)
فتمرّ : فيمرّ (ي)
[١٥٦] واما الذي : والذين (ي)
فليست : فليس (ي ، ك)
تكون : يكون (ي)
[١٥٧] ميل : ميول (ي ، ك)
وتسمّي هذه : ويسمّي (ي)
تخرج : يخرج (ي)
[١٥٨] لدائرة : الدائرة (ي)
تكون : يكون (ك)
حركته : حركتها (ي)
الى : والى (ي)
[١٦٠] السمتية : الشمسية (ك)
[١٦١] وتنتهي : وينتهي (ي ، ك)
[١٦٢] فتكون : فيكون (ي ، ك)
[١٦٢] تصغر : يصغر (ي ، ك)
وكلّما : وكلّ ما (ي)

جهاز التحقيق

[١٦٢] الطالعة : + الطالقة لها (ي)
[١٦٤] تتحرّك : يتحرك (ي)
وتكون : ويكون (ي)
تتحرّك : يتحرك (ي)
تنتهي : ينتهي (ي)
فتنتهي : فينتهي (ي)
النهار : + لانّ الدائرة نصف النهار (ي)
تقطع : يقطع (ي)
يتحرّك : تتحرّك (ي ،ك)
[١٦٥] جيب : أجيب من (ي)
وتنتهي : وينتهي (ي)
[١٦٦] الجيب : الخيب (ي)
[١٦٧] المعكوس : المنكوس (ك)
تستعمل : يستعمل (ي ، ك)
تستخرج : يستخرج (ي ، ك)
[١٦٩] اذا : كلما (ي)
[١٧٠] المتوازية : الموازية (ي؛ وهذه الكلمة غير واضحة في المخطوط)
في : من (ي؛ ونجد كلمة « في » مكتوبة فوق كلمة « من »)
ليست : ليس (ي ، ك)
[١٧١] فاما : + حال (ك)
فهو: فهي (ي ، ك)
تطلع : يطلع (ي)
[١٧٢] ونمثّل : ويمثل (ك) ؛ ونمتل (ي)
تكون : يكون (ي ، ك)
من : بين (ك)
فتتحرّك : وتحرك (ي)
ما : _ (ك)
تتحرّك : يتحرك (ي ، ك)
تكون قد طلعت : يكون قد طلع (ي ، ك؛ وفي مخطوط « ك » نجد « قد طلع » في الهامش)
فتكون : فيكون (ي ، ك)
القوس : + الكوكب (شطب في مخطوط « ك »)
دائرة : دائر (ي)
[١٧٣] الدائرة : + البروج هذه (شطب في مخطوط « ك »)

جهاز التحقيق

[١٧٢] مسامتتها : مسامتها (ي)
دائرة : داير (ي) ؛ وفي مخطوط « ك » نجد « فلك » فوق « دائرة »
[١٧٣] متساوية : متساويان (ي)
[١٧٤] فلكلّ : لكل (ي)
معدّل : ممطالع (ي)
ومطالعات : ومطالع (ي)
درجات : الدرجات (ي)
[١٧٥] ايضاً : ـ (ك)
فانّ : ـ (ك)
تساوي مطالعه : يساوي مطالعة (ي)
مطالعه : مطالعة (ي)
النظير : النظيرة (ي)
[١٧٦] مخالفة : مخالف (ي ، ك)
[١٧٧] معدّل : معدّ (ي)
[١٧٨] اطول : طول (ي)
للافق : لافق (ي)
هذا : هذه (ي)
تنطبق : ينطبق (ي)
وليله : وليلة (ي)
[١٧٩] آن : آان (ي)
[١٨١] ظاهراً : طالعا (ي)
دائرة : ـ (ي)
الجزء الظاهر...الخفي : الجزين الظاهر والخفي (ي) ؛ الجزء وبين الظاهر والخفي (ك)
[١٨٢] متّصلتين : متصلين (ي ، ك)
[١٨٥] سقط من مخطوط « ك» هذه الفقرة والفصلان التاليان الخاصان بالشمس والقمر
فلكاً : فلك (ي)
يخصّه : تخصّه (ي)
باشرفها : باشرافها (ي)
نتبعه : يتبعه (ي)
[١٨٧] والسطح : السطح (ي)
منها : منهما (ي)
من الكرة : مركزه (ي)
الفلك : ـ (ي)

جهاز التحقيق

[١٨٧] يتحرّك : تتحرك (ي)
ومحوره : وامحوره (!)(ي)
[١٨٨] مشتركتين : مشتركين (ي)
قطبيه : قطيبه (ي)
[١٩٠] تسمّي : يسمي (ي)
منهما : منها (ي)
[١٩١] تكون : يكون (ي)
وتكون : ويكون (ي)
[١٩٢] تسمّي : يسمي (ي)
[١٩٢] احديهما : احدهما (ي)
نقطتي : نقطة (ي)
تخرج : يخرج (ي)
محيط: الكرة (ي)
تخرج : يخرج (ي)
[١٩٤] محيط : غير واضحة في المخطوط
تسمّي : يسمي (ي)
وتسمّي : ويسمي (ي)
[١٩٥] في كتابه : غير واضحة في المخطوط
[١٩٦] وهذه : يلي هذه الكلمة حرفان غير واضحان في المخطوط
منتقلة : مننقلة (ي)
للانقلاب : المنقلب (ي)
[١٩٩] تتحرّك : يتحرك (ي)
يسامت : تسامت (ي)
تخرج : يخرج (ي)
الى : ــ (ي)
وتنتهي : وينتهي (ي)
تنتهي : ينتهي (ي)
[٢٠٠] تكون : يكون (ي)
تزال تقرب : يزال تقرب (ي)
تنتهي : ينتهي (ي)
تكون : يكون (ي)
تزال : يزال (ي)
تعود : يعود (ي)

جهاز التحقيق

[٢٠١] تقطع : يقطع (ي)
[٢٠٢] تكون : يكون (ي)
ذكرها : ــ (ي)
تكون : يكون (ي)
[٢٠٤] حضيضها : خضيضها (ي)
[٢٠٥] تحدث : يحدث (ي)
وليست : وليس (ي)
[٢٠٦] ليس : ليست (ي)
تكون : يكون (ي)
القطعة : غير واضحة في المخطوطة
فتكون : فيكون (ي)
[٢٠٧] ليس : ليست (ي)
تكون : يكون (ي)
[٢٠٨] وتسمّي : ويسمي (ي)
الاستواء : الاستوى (ي)
لقربه إ القربه (ي)
[٢٠٩] ولنخطّ : وليخط (ي)
تحدث : يحدث (ي)
تحدثها : تحدث بحدوثها (ي)
[٢١١] السطح : سطح (ي)
منهما : منه (ي)
[٢١٥] الادنى : لادنى (ي)
[٢١٧] مشتركتين : مستركين (ي)
تتحرّك : يتحرك (ي)
تسمّي : يسمي (ي)
[٢٢٠] تحرّك : حرك (ي)
متّصل : متضل (!)(ي)
[٢٢١] يرسم : ترسم (ي)
[٢٢٢] نقطتين : نقطتي (ي)
مركزها : مركزهما (ي)
[٢٢٣] تسمّي : يسمي (ي)
[٢٢٧] تسمّيان : يسميان (ي)
من : ــ (ي)

جهاز التحقيق

[٢٢٧] فتسمّي : فيسمي (ي)
[٢٢٨] تلي: يلي (ي)
[٢٣٠] فيحرّك : فتحرك (ي)
التي فيه : + وحركته على خلاف توالي البروج (ي)
[٢٣٢] فيحرّك : فتحرك (ي)
وتكون : ويكون (ي)
[٢٣٦] تكون : يكون (ي)
[٢٣٧] وانتقل : وينتقل (ي)
عن : على (ي)
[٢٣٩] مختلفتين : مختلفين (ي)
[٢٤٠] تتحرّك : يتحرك (ي)
فتنتقل : فينتقل (ي)
تعود : يعود (ي)
[٢٤١] مختلفتين : مختلفين (ي)
[٢٤٢] مقابلاً : تقاطرا (ي)
تكون : يكون (ي)
فتكونان : فيكون (ي)
[٢٤٥] فليست تكون : فليس يكون (ي)
خلاف : غير واضح في المخطوط
فتكون : فيكون (ي)
وتكون : ويكون (ي)
[٢٤٦] وعلى : ـ (ي)
وتسمّي : ويسمي (ي)
المحرّك : المحركة (ي)
وتسمّي : ويسمي (ي)

جهاز التحقيق

[٢٤٧] فلنتوهّم : فلتوهم (ي)
تخرج : يخرج (ي)
[٢٤٨] تقطع : يقطع (ي)
هي : هو (ي)
[٢٤٩] فانتقل : انتقل (ي)
[٢٥٠] فتتحرّك : فيتحرك (ي)
نقطتها : نقطته (ي)
فتكون : ويكون (ي)
[٢٥١] تتغيّر : يتغير (ي)
فتارةً...تنقص : فتزيد فتارة وتنقص اخرى (ي)
[٢٥٢] وللدائرة : ولدائرة (ي)
[٢٥٦] عنه : غير واضحة في المخطوطة
[٢٥٧] بالاسفار : بالاستفار (ي)
فتبتدئ: يبتدى (ي)
بالاسفار : بالاستفار (ي)
[٢٥٨] فيصير : غير واضحة في المخطوطة
[٢٥٩] واستنار : غير واضحة في المخطوطة
[٢٦٠] تلتقي : + فيتشكل الشعاع الذي يخرج من محيط جرم الشمس الى جرم الارض ويلتقي (ي)
[٢٦١] ويعود : ود (!) (ي)
[٢٦٥] ان : لم (ي)
بعضه : بعضها (ي)
قريبة : قريب (ي)
[٢٦٦] فنرى : فترى (ي)
[٢٦٧] مكاننا : مكانا (ي)
[٢٦٩] مليئاً : ما يلا (ي)
تحيط : يحيط (ي)
بالجزء : الخزء بالنصف (ي)
[٢٧٠] وآخره : وآخراه (ي)
[٢٧٢] فهذه : فهذا (ي)
ولكي : ولكن (ي)
واضحاً : واضجا (!)(ي)
[٢٧٣] مستدير : مستديرة (ي)

[٢٧٦] وتنفصل : وينفصل (ي)
ويفرزها : وهزرها (ك)
الفلك : فلك (ي)
مشتركة : مستركة (ي)
فلك : ـ (ي)
للفلك الحامل : للحامل (ي)
[٢٧٧] وتنفصل : وينفصل (ي , ك)
ويسمّي هذا الفلك...الحامل : ويسمّي هذا الفلك الخارج المركز الحامل (ي)
على : ـ (ك)
توالي البروج : في الهامش في مخطوط « ك »
[٢٧٨] وتنفصل : وينفصل (ي)
وتنفصل...مستديرة : في الهامش في مخطوط « ك »
الثابتين : ثابتين (ك)
مشتركتين : ـ (ك) ؛ مشتركين (ي)
على نقطتين...المتوازيين: في الهامش في مخطوط « ك »
وتسمّي : ويسمّي (ي)
[٢٨٠] حركة : حركت (ك)
[٢٨١] متوازيتين : متوازيين (ي)
[٢٨٢] ويحدث : وحدث (ك)
سطحي : سطح (ي , ك)
الاول الذي : غير واضحة في مخطوط « ي »
وتسمّي : ويسمّي (ي)
[٢٨٣] دائرة : غير واضحة في مخطوط « ي »
مقاطعة : متقاطعة (ي)
تسميان : يسميان (ي , ك)
الجوزهرين : جوزهرين (ي)
وتسمّي : ويسمّي (ي)
[٢٨٤] ولنتوهم : وليتوهم (ك)
التي : الذي (ي)
فطرف : وطرف (ي)
وهو : فهو (ي)
ومقابله : ومقابلة (ي)

جهاز التحقيق

[٢٨٥] فيحرك : فتحرك (ك)
نقطتي : نقطيي (!)(ي)
[٢٨٦] جزء واحد : جزا واحد (ي) ؛ جزؤ واحد (ك)
[٢٨٧] ثابتين : الثابتين (ي)
فيحرك : فتحرك (ك)
ويحرك : وتحرك (ك)
[٢٩١] فيحرك : فتحرك (ك)
[٢٩٢] وتحدث : ويحدث (ي)
تسمي : يسمي (ي) ؛ ويسمي (ك)
[٢٩٢] حركته : غير واضح في مخطوط « ي »
تكون : يكون (ي)
ادناها : اذياها (ي)
[٢٩٤] ويتغير : وتتغير (ي)
[٢٩٥] فنتوهم : فيتوهم (ك)
فيسمي : يسمي (ي)
لقطر : لقاطر (ي)
قطره المذكور : قطرة المذكور (ي)
تسمي : يسمي (ي)
يحدها : تحدها (ك)
[٢٩٦] فتكون : فيكون (ي ، ك)
الثلاثة : الثلث (ي ، ك)
التي تلي : الذي يلي (ي)
مركز : ــ (ي، ك)
يدور : تدور (ي)
يفارقه : غير واضح في مخطوط « ي »

هاز التحقيق

٢٩٨] تكون : يكون (ي ، ك)
٢٩٩] مركز : مركزي (ي)
٣٠٠] تكون : يكون (ي)
معا : ـــ (ك)
٣٠] ويكون : وتكون (ك)
٣٠٢] ويقال...ايضا: ونقلانه نقلا به ايضا يكون (ي)
وذلك : وذاك (ي)
٣٠] وحركة الكوكب...التدوير: ـــ (ي) ؛ في الهامش في مخطوط « ك »
انما يتباعد : ايضا بتباعد (ي)
اعلى : اعلا (ك)
تدويره : التدوير (ي)
متساويا : مساويا (ي)
٣٠] فلك : ـــ (ي)
فيسامت : فتسامت (ك)
تسبق : يسبق (ي)
٣٠] فحركته : وحركته (ي ، ك)
بحركة : بحركته بحركته (!)(ي)
فحركته : وحركته (ي ، ك)
قوسا : غير واضح في مخطوط « ي »
على طريق المثال : + مثالا (في الهامش في مخطوط ي)
جزء واحد : جزءا واحدا (ي)
بمقدار : لمقدار (ك)
فيكون : غير واضحة في مخطوطة « ي »
٣٠] باق : بباق (!) (ك)
٣٠] طرفه : + الاقرب (ك)
وتسمي : تسمي (ي)

[٢١٠] المقاطع : + ايضا (ي)
لوضع : لموضع (ي)
مجاز الكوكب : مجازا لكوكب (ي)
[٢١٢] فلك : ـ (ي)
[٢١٥] الفلك : فلك (ي)
[٢١٦] فاما : واما (ي)
بالفارقة : بالفارقة (ي)
الرأس : ـ (ك)
وهي الجوزهر : ـ (ي)
فيتحرك : فتتحرك (ي)
[٢١٨] وحركتها : + في الطول (ك)
[٢١٩] هذه الحركات مرقومة بحروف أبجد في مخطوط ي عدا واحدة منها.
وتسمي : ويسمي (ي)
المدير : المديره (ي)
تكون : يكون (ي)
حركة الطول : وهذه الحركة هي التي لم ترقم في مخطوط « ي »
اعني...الاقرب : ـ (ك)
[٢٢٠] تقطع : يقطع (ي , ك)
تسمي : يسمي (ي)
[٢٢٢] منهما : ـ (ك)
سطحه : سطح (ك)
[٢٢٢] الكرة : غير واضح في مخطوط « ي »
مستديرة : مستدير (ي)
وهي تتحرك : وهو يتحرك (ي , ك)
لها : له (ي , ك)
مستوية : في مخطوط « ك » شطبت كلمة « مستديرة »
[٢٢٤] سطحيها : سطحيه (ي , ك)
تتحرك : يتحرك (ي)
مركوز : مركوزة (ي)

جهاز التحقيق

[٢٢٥] حدثت : حدث (ي ، ك)
ونتوهم : ـ (ك)
نتحدث : فحدث (ي)
[٢٢٦] ونتوهم : ويتوهم (ك)
مستقيما : ـ (ي) ؛ في الهامش في مخطوط « ك »
الابعد : + الابعد (ي)
[٢٢٧] ونتوهم : ويتوهم (ك)
[٢٢٨] وتسمي : ويسمي (ي)
[٢٣٠] الخاصين : الخاصيين (ي)
مستديرة : مستدير (ي)
ويحرك : وتحرك (ي ، ك)
ايضا : ـ (ي)
[٢٣١] الاولى : الاولة (ي)
[٢٣٢] تسمي : يسمي (ي)
[٢٣٣] يقود : غير واضح في مخطوط « ي »
كمثل : لمثل (ك)
تكون : يكون (ي)
عن : على (ي)
وهي : هي (ي)
والمفارقة التي تكون عن سطح : شطب في مخطوط « ك »
احدي : اخري (ي) ؛ احد (ك)
لها الذنب : الذنب لها (ي) ؛ لها (ك)
[٢٣٤] مركزهما : مركزها (ي)
لكن : + من (ي)
وعودات : وعددات (ي)

جهاز التحقيق

[٢٢٥] مركز : ـ (ك)
فلك : ـ (ك)
[٢٢٦] ترقم هذه الحركات في هامش مخطوط « ي »
فلكه : فلكها (ي)
والحركة : وحركة (ي)
قطر : فطر (!)(ي)
[٢٢٦] المنحرف : المنحرفه (ك)
[٢٢٧] هذا : هذه (ي)
فجميع : غير واضح في مخطوط « ك »
نورده : يورده (ي)
مشروح : مشروج (ك)
[٢٢٨] والمشتري : والمستري (ي)
في التعليم : + في التعليم (هكذا في ك)
ولذلك لما : + ولذلك لما (هكذا في ك)
تختلف : يختلف (ي)
[٢٢٩] تخصه : تخصها (ي)
وتحيط : ويحيط (ي)
[٢٤٠] سطحه : وسطحه (ك)
يماس : مماس (ي)
محدب : ـ (ي)
يماس : ـ (ي)
الاعلى : يماس (ي) ؛ الاعلا (ك)
فلك : ـ (ي)
[٢٤٢] وتنفصل : وينفصل (ك)
بها : به (ي)
تتحرك : يتحرك (ي)
مستوية : غير واضح في مخطوط « ي »
وتسمي : ويسمي (ي)
[٢٤٣] هذا : هذه (ك)
تسمي : يسمي (ي ، ك)
واحدة : واحد (ي)
تتحرك : يتحرك (ي)
مركزها : مراكزها (ي)
[٢٤٣] اقطابها : اقطادها (!)(ي) جهاز التحقيق

[٢٤٤] وجرم : جرم (ي)
جسم : جرم (ي)
[٢٤٥] تخصه : يخصه (ي)
تسمي : يسمي (ي)
واحدة : واحد (ي)
تنتقل : ينتقل (ي ، ك)
تتغير : يتغير (ي ، ك)
[٢٤٦] ونتوهم : ويتوهم (ك)
استقامته : استقامة (ي)
فتكون : ميكون (!)(ي)
[٢٤٧] واحد : واحدة (ك)
ونقطة البعد الاقرب...الجنوب : ــ (ك)
[٢٤٨] ذلك : وذلك (ي)
مركز : مركزي (ي)
[٢٥٠] كل واحد من : ــ (ي)
لقطبي : لمقطبي (هكذا في ك)
ويتحرك : وتحرك (ك)
يحيط : تحيط (ك)
وينقل : وينتقل (ك)
[٢٥١] وينتقل مع جملة...جرم : وجرم (ك)
[٢٥٢] فتحدث : فيحدث (ي)
تسمي : يسمي (ي ، ك)
[٢٥٣] فعلى : ففي (ك)
[٢٥٥] الفلك : ــ (ي)
تكون : يكون (ي ، ك)
اعني الرأس : ــ (ي) ؛ فوق السطر في مخطوط « ك »
[٢٥٦] الحال : ــ (ك)
[٢٥٧] مطابقة : مطابقه (ي)
[٢٥٨] هذه الحركات مرقومة في مخطوط « ي » بزيادة واحدة.
الطول : + وحركة مركز الكوكب حول مركز فلك التدوير (ي)

[٢٥٩] فهذا : غير واضح في مخطوط « ك »
اقتنعنا : اقنعنا (ك)
بالايجاز : في الايجاز (ي)
الكواكب : الكوكب (ي)
[٢٦٠] والسطح : وسطح (ي)
[٢٦١] على : الى (ك)
الاجزاء : اجزاء (ي)
هما : _ (ي)
وجده : وجد (ي)
[٢٦٢] تتغير : يتغير (ي ، ك)
تنتقل : انتقل (ي ، ك)
فتتحرك : فيتحرك (ي ، ك)
[٢٦٣] منها : منهما (ي)
الفلك : + جميع الكواكب الثابتة وكل كوكب منها يحدث (ي)
وهذه : وهذا (ي)
[٢٦٤] وتقطع : ويقطع (ي)
[٢٦٥] هي : هو (ي)
المتحركة : المتحرك (ي)
هي : فوق السطر في مخطوط « ي »
[٢٦٧] هي : هو (ك) ؛ _ (ي)
[٢٦٨] الكواكب : الكوكب (ي)
بست : بسته (ي)
تخرج : يخرج (ي ، ك)
فتقسمه : فنقسمه (ي) ؛ فيقسمه (ك)
فيما : بما (ك)
الست : الستة (ي)
[٢٦٩] وتقسم : ويقسم (ي ، ك)
الاثنى : الاثنا (ك)
تسمي : يسمي (ي)

جهاز التحقيق

[٢٧٠] فيه : فيها (ي)
وتسمي : ويسمي (ي)
باسماء : باسم (ي)
باسماء : باسم (ي ، ك)
باسماء : باسم (ي)
جميع الفلك الاعلى...فاقسام : ــ (ك)
تسمي : يسمي (ي)
باسماء : باسم (ي ، ك)
[٢٧١] رصدت : صدت (!)(ي)
من : في (ي)
جمع : جميع (ي)
[٢٧٢] تسعة : ثمانية (ك)
والخفية : والحفية (ي)
خمسة : خمس (ي)
[٢٧٣] ايضا : ــ (ي)
مواضعها : مواضها في (ي)
فجمع : فجميع (ي)
واربعين : واربعون (ي)
يسمي : سمى (ك)
[٢٧٤] فهو : وهو (ك)
الذي : + هو (ي)
لعلة قربه : لعله قريب (ي ، ك)
اهل : ــ (ك)
فلك : الفلك (ك) ؛ ــ (ي)
الثابتة : + الشكل لا الكواكب الثابتة. صورة افلاك الكواكب الثابتة (ك)
[٢٧٥] اما : فاما (ك)
[٢٧٦] المائلة : ــ (ي)
توجد منها : منها يوجد (ي)
اليها : ــ (ي)
تقاس : يقاس (ي ، ك)
تخرج : يخرج (ي)
بمراكز : بمركز (ي)

جهاز التحقيق

[٢٧٨] بالحقيقة : لحقيقة (ك)
الاعلى : الاعلا (ك)
[٢٧٩] واما : فاما (ي)
موضعهما : موضعهاما (!)(ي)
بموضعهما : بموضعها (ي)
[٢٨٠] اراء : اراآ (ي)
وعلى : على (ك)
بالاستقراء : بالاستقرار (ي)
ينتهي : انتهي (ي)
على : وعلى (ي)
تدمنا : ـ (ي)
[٢٨١] وثماني : وثمنية (ي)
للكواكب : الكواكب (ي)
وثماني : وثمان (ي)
فلك : الفلك (ي)
[٢٨٢] وتخطيط : وتحيطنا (ي)
مفردة : مقردة (!)(ي)
وعلى : على (ك)
متفارقا : مفرّقا (ك)
جملتها : جميعها (ي)
[٢٨٣] فيستعجم : ويستعجم (ي)
نقتصر : تقتصر (ي)
[٢٨٤] لجميع : بجميع (ك)
وابعد : ـ (ك)
[٢٨٥] التداوير : التدواير (ي) ؛ وكذا شطب في مخطوطة « ك »
وكوكب : وكواكب (ي)
ونجعله تريبا من مركز العالم ليكون ابين : ـ (ي)
وهذه : ـ (ك)

جهاز التحقيق

آخر مخطوط « ي »

والشكر : ولشكر

التعليق

يخل : يحمل (شرام)
تتحرك : يتحرك
منها : منهما
يوجد : توجد (شرام)
منقادة : منقاد
تتحرك : يتحرك
تحتاج : يحتاج

فصل آخر

افلاك : الافلاك (هكذا في المخطوط وعند شرام ايضا)
حركة : حركات (شرام)

خاتمة مخطوط « ك »

على: ـ
نسخت : نسخ
منها : منه
المذكورة : المذكور